Satellite Technology

ELECTRONIC MEDIA GUIDES

Elizabeth Czech-Beckerman   *Managing Electronic Media*

Ken Dancyger   *Broadcast Writing: Dramas, Comedies, and Documentaries*

Donna L. Halper   *Full-Service Radio: Programming for the Community*

Donna L. Halper   *Radio Music Directing*

William Hawes   *Television Performing: News and Information*

Robert L. Hilliard   *The Federal Communications Commission: A Primer*

John R. Hitchcock   *Sportscasting*

Andrew F. Inglis   *Satellite Technology : An Introduction*

Carla B. Johnston   *Election Coverage: Blueprint for Broadcasters*

Marilyn J. Matelski   *TV News Ethics*

Marilyn J. Matelski   *Daytime Television Programming*

Robert B. Musburger   *Electronic News Gathering: A Guide to ENG*

Satellite Technology
An Introduction

Andrew F. Inglis

Focal Press
Boston   London

FEB 2 6 1992

AAY 7166

Library of Congress Cataloging-in-Publication Data
Inglis, Andrew F.
　Satellite technology: an introduction / Andrew F. Inglis
　　　p.　　　cm.—(Electronic media guide)
　Includes bibliographical references and index.
　ISBN 0-240-80078-8
　1. Artificial satellites in telecommunication.　2. Earth stations
　(Satellite telecommunication)　I. Title. II. Series.
　TK5104.I53　1991
　621.382'54—dc20
　　　　　　　　　　　　　　　　　　　90-45629
　　　　　　　　　　　　　　　　　　　CIP

British Library Cataloguing in Publication Data
Inglis, Andrew
　Satellite technology: an introduction.–(Electronic media guide).
　1. Broadcasting. Use of communications satellites.
　I. Title　　　II. Series
　621.3897

　ISBN 0-240-80078-8

Butterworth–Heinemann
80 Montvale Avenue
Stoneham, MA 02180

10　9　8　7　6　5　4　3　2　1

Printed in the United States of America

Contents

Preface

During the past 15 years there has been a phenomenal growth in the use of communication satellites for the transmission and distribution of radio and television programs. Satellites have evolved from a novelty in high technology with an uncertain future, to an indispensable component of these industries. They were initially perceived to be useful mainly for the transmission of voice and data traffic, but as their capabilities became better understood they began to play an equal, if not more important role, in the transmission and distribution of television programs. Later, they became widely used for the distribution of radio programs.

There is now a synergistic relationship between satellites and the radio and television industries. These industries provide a major market for satellite communication services but also are highly dependent on them. Neither cable television (CATV) nor electronic news gathering (ENG) could have reached its present state of development without the use of satellites.

The introduction of satellites into the radio and television industries created a new set of challenges and opportunities for the members of their technical communities. The new concepts, new vocabulary, and new technologies of satellites must be learned by the engineers responsible for the design and operation of broadcasting and cable TV systems. It is equally a requirement for students who are pursuing a course of study leading to employment in the engineering departments of broadcasting and cable TV companies.

This volume contributes to the learning process by providing an introduction to satellite technology in language that is accessible to those who are not specialists. The scope of its subject matter is broad, ranging from the theory of satellite operation to practical instructions for the initial setup of mobile earth stations. For those who wish to pursue the study of the technical aspects of satellites further, a comprehensive list of references are included.

The author is indebted to the many members of the broadcasting and satellite industries who supplied reference materials and invaluable advice and suggestions. He is grateful to Wayne Rawlings and his staff from Station KCRA-TV in Sacramento who provided information on the practical operation of satellite news gathering (SNG) earth stations. Most of all he wishes to thank Walter Braun, John Christopher, and Marvin Freeling of GE American Communications. They were an indispensable resource, and they carefully reviewed the manuscript for technical accuracy. Any remaining errors, however, are the author's!

1

▼
▼
▼
▼
▼

SATELLITE COMMUNICATION SYSTEMS

OVERVIEW

This chapter is an introduction to the technology of the satellite communication systems used for the transmission and distribution of radio and television programs. Succeeding chapters describe the applications and technical characteristics of satellites and earth stations, earth station equipment and design procedures, Federal Communications Commission (FCC) rules and regulations, earth station installation and operational procedures, and sources of satellite services and equipment.

Satellite Communication System Elements

The elements of a satellite communication system are shown in Figure 1.

The signal to be transmitted, that is, the *baseband signal,* is received at an *earth station,* where it modulates a high-power radio frequency transmitter. The earth station antenna radiates the transmitter signal to a *geosynchronous* satellite, which appears to remain in a fixed position in space. The satellite receives the radiated signal, shifts its frequency, and amplifies it by means of a *transponder*, then reradiates it to back to earth where it can be received by earth stations in the coverage area of the satellite.

Earth stations form the *ground segment* of a satellite communication system, while the satellite constitutes the *space segment*. The transmission system from earth station to satellite is called an *uplink*, and the system from satellite to earth is called a *downlink*.

Satellite Service Areas, or *Footprints*

Earth station antennas have very narrow beams, both to increase their gain and to avoid interference with adjacent satellites. By contrast, the antennas on communication satellites usually have rather broad beams so that they can provide service to and from a large area—typically an entire region, country, or even an entire hemisphere. In an exception to this practice, some satellites have narrow *spot beams* for specialized service to limited areas. The area which receives a signal of useful strength from the satellite is known as its *footprint*.

Satellite Frequency Bands

Three frequency bands, C-band, Ku-band, and DBS (direct broadcast by satellite) are used for satellite transmission of radio and television programs. C-band downlink transmissions are in the 4 GHz region of the spectrum, which is shared

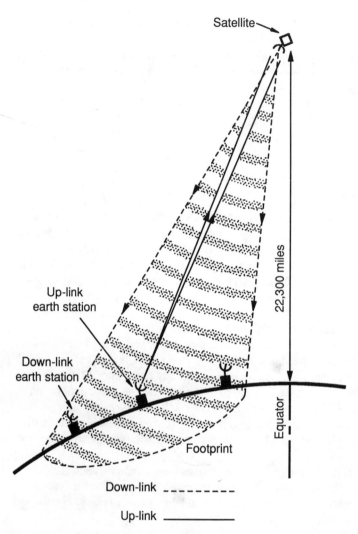

▶ *Figure 1 Satellite communication system.*

with terrestrial microwave services. Ku-band downlinks operate in the 12 GHz region and have exclusive use of these frequencies. Direct Broadcast Satellite (DBS) downlinks operate in the 12.5 GHz region and are intended for direct pickup by small home antennas. The technical characteristics of these frequency bands are described in Chapter 3.

Satellite Transmission Modes

Frequency modulation is ideally suited for the transmission of television signals, and it has been almost universally used to date. It has three important advantages over amplitude modulation for this purpose:

1. It does not require highly linear power amplifiers, either in the satellite or the uplink.
2. It has a substantial *noise improvement factor* (see Chapters 4 and 5), so that the signal-to-noise-ratio of the output video signal is higher than that of the radio frequency carrier.
3. The transmitted energy can be more uniformly distributed across the channel bandwidth by *sideband energy dispersal*. This process is important for C-band systems because it increases the legally permissible downlink power (see Chapter 4).

The other transmission mode alternative, digital transmission, is not presently used for television signals because of its bandwidth requirements. It is, however, excellently adapted for high-fidelity audio signals, and it is widely used by radio networks for program distribution (see Chapters 2 and 5).

Somewhat paradoxically, the digital mode also has the potential for *reducing* the bandwidth required for a video signal transmission system. This results from the repetitiveness and large amount of redundancy in the video signal and requires advanced digital processing techniques. As of this writing (1990), none of them has had the benefit of extensive tests. Signal processing for bandwidth reduction invariably results in at least a small reduction in picture quality, such as, reduced sharpness in the images of moving objects.

Competitive Transmission Mediums

Satellites compete with other communication mediums—microwave, coaxial cable, and fiber optic cable—for the transmission of television signals. Each of these mediums has characteristics which make it especially suited for certain types of service. None of them excels in all respects, and all will continue to be used extensively in the foreseeable future.

To date, the competition to satellites for television service has come primarily from coaxial cable and microwave. Fiber optics is now coming into use and will significantly change the competitive situation in the future.

Satellites and Fiber Optics

The most striking feature of fiber optic systems is their enormous bandwidth—a typical system has a bandwidth of 3 GHz (3,000 MHz) as compared with 36 MHz for a satellite transponder or 864 MHz for the 24 transponders in a C-band satellite.

This bandwidth makes it possible for fiber optic systems to transmit video signals in a digital rather than an analogue format. The digital format has many advantages, but its huge bandwidth requirements have made the medium and long-haul digital transmission of video signals difficult or impractical with coaxial cable, microwave, or satellites (but note the earlier discussion of possible future bandwidth reduction by the use of digital transmission).

Fiber optic cable systems are now being installed for intercity, point-to-point transmission circuits by all major communication carriers for voice, data, and video. These facilities will doubtless provide very serious competition to satellites for fixed, point-to-point video transmission services.

In addition, long-range planners in the communications industry have envisioned a future for fiber systems that goes beyond fixed, point-to-point services. Their proposal is to provide every home and business with a fiber optic cable connection to a nationwide communication system that would use a technology known as ISDN (integrated services digital network), or its equivalent. Two-way digital bit streams on the cable would provide radio and television programs, telephone service, facsimile service, and access to remotely located computers and a variety of data banks to every home and business.

This is an extremely attractive concept, and if it proves to be technically feasible, economically competitive, and politically possible, it could make satellites obsolete for most television communication services, except mobile applications. (It could also make television broadcasting and present-day cable TV obsolete.) The installation and operation of such a nationwide system, however, involves massive technical, financial, and political problems, not all of which have clear solutions. At best, it will be more than a decade before such a system can be designed, approved by the government, financed, constructed, and put into operation. In the meantime, satellites will continue to play a vital role in the television industry.

THE GEOSYNCHRONOUS ORBIT

Location

Geosynchronous satellites are located in the *geosynchronous orbit,* which forms a circle in the plane of the equator, 22,300 miles above the earth (see Figure 2). They revolve once each day in synchronism with the earth's rotation; and at this elevation the gravitational force pulling them toward the earth is exactly balanced by the centrifugal force pulling them outward. Since they revolve at the same rotational speed as the earth, they appear stationary from the earth's surface; and radio signals can be transmitted to and from them with highly directional antennas pointed in a fixed direction. It is this property that makes satellite communications practical.

Orbital Slots

International regulatory bodies and national governmental organizations, such as the Federal Communications Commission (FCC) (see Chapter 6), designate the locations on the geosynchronous orbit where communication satellites can be located. These locations are specified in degrees of longitude and are known as *orbital slots.*

Since all communication satellites operate in the same frequency bands, the spacing between orbital slots must be great enough to reduce to an acceptable level the interference between transmissions to and from adjacent satellites. The minimum spacing required to achieve this objective depends on the width of the earth station antenna beams, that is, the directivity of their antennas.

In response to the huge demand for orbital slots, the FCC has progressively reduced the required spacing and has established a future standard of only 2° for C- and Ku-band satellites—a spacing which requires the use of very narrow antenna beams. DBS satellites are designed to be received by smaller and less directional receiving antennas in individual homes (see Chapter 4 for the relation between antenna size and beam width). Accordingly, DBS orbital slot assignments are located at intervals of 9°.

Western Hemisphere Orbital Slot Allocations and Assignments

By international agreement negotiated through the International Telecommunications Union (ITU), each country is allocated an arc of the geosynchronous orbit, within which it can assign orbital slots. A national regulatory body—the FCC in the case of the United States—makes the slot assignments within this arc. The current (1990) orbital slot allocations and assignments for the western hemisphere are tabulated on the following pages (see Chapter 3 for the definitions of fixed service satellites (FSS), direct broadcast satellites, C band and Ku band). The slot assignments change frequently, and the tables must be updated periodically.

United States   The United States is allocated the orbital arcs 62°-103° and 120°-146° west longitude for C-band satellites and 62°-105° and 120°-136° west longitude for Ku-band satellites. Within each band, a single satellite is assigned to each slot. The assignments as of January 1990 within these arcs are shown in Table 1 and Figure 3.

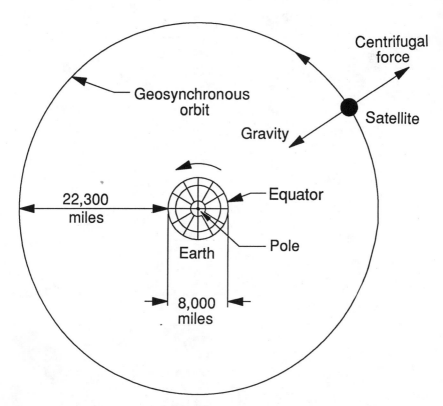

▶   *Figure 2 Geosynchronous satellites.*

▶ **Table 1**   United States orbital slot assignments, January 1990

C-band Fixed Service Satellites (FSS)		
W. Long.	*Name*	*Operator*
69°	Spacenet-2 (hybrid) H[1]	GTE Spacenet
72°	Satcom-2R V	GE Americom
74°	Galaxy-2 H	Hughes
76°	Comstar-4 V	Comsat
79°	Satcom H-1 (Hybrid) H	GE Americom
81°	Satcom-4/Galaxy-5E V	GE/Hughes
83°	ASC-2 H	Contel ASC
85°	Telestar-302 V	AT&T
87°	Spacenet-3 (hybrid) H	GTE Spacenet
91°	Galaxy-6 H	Hughes
93°	Spotnet-1 (hybrid) V	Spotnet
95°	Galaxy-3/3R H	Hughes
97°	Telstar-301/401(hybrid) V	AT&T
99°	Westar-4/Galaxy-4R H	Hughes
101°	Contelsat-1 (hybrid) V	Contel ASC
103°	Spacenet-1R (hybrid) H	GTE Spacenet
123°	Telstar-303 V	AT&T
125°	Galaxy-5W H	Hughes
127°	Spotnet-2 (hybrid)	Spotnet
129°	ASC-1 (hybrid) H	Contel ASC
131°	Satcom-3R V	GE Americom
133°	Galaxy-1/1R H	Hughes
135°	Satcom-C4 V	GE Americom
137°	Satcom-1R/C1 H	GE Americom
139°	Aurora-2 V	Alascom
143°	Aurora-1	Alascom

Ku-band Fixed Service Satellites (FSS)		
62°	AMSC	AMSC
69°	Spacenet-2 (hybrid)	GTE Spacenet
72°	SBS-6	Hughes
74°	SBS-1/2	Hughes
79°	Satcom-H1 (hybrid)	GE Americom
81°	Satcom-K2	GE Americom
85°	Satcom-K1	GE Americom
87°	Spacenet-3 (hybrid)	GTE Spacenet
91°	SBS-4	IBM
93°	Spotnet-1 (hybrid)	Spotnet

▶ **Table 1** *(continued)*

W. Long.	Name	Operator
95°	SBS-3	MCI
97°	Telstar-301/401 (hybrid)	AT&T
99°	Galaxy A-R	Hughes
101°	Contelsat-1 (hybrid)	Contel-ASC
103°	Spacenet-1R (hybrid)	GTE Spacenet
105°	GStar-2	GTE Spacenet
121°	GStar-1/1R	GTE Spacenet
123°	SBS-5	IBM
125°	GStar-4	GTE Spacenet
127°	Spotnet-2 (hybrid)	Spotnet
129°	ASC-1 (hybrid)	Contel-ASC
131°	Galaxy-BR	Hughes
139°	AMSC	AMSC

Ku-band Direct Broadcast Satellites (DBS)

61.5°		
101.0°		
110.0°	See Chapter 2 for channel	
119.0°	assignments within these	
148.0°	slots.	
157.0°		
166.0°		
175.0°		

[1] V(ertical) and H(orizontal) following the satellite's name denotes the polarity on the downlink of the odd-numbered C-band transponders (see Chapter 3).

Source: *Broadcasting/Cable Yearbook.* Washington, D.C.: Broadcast Publications, Inc., 1990.

The United States is also allocated the eight DBS slots in the arc 62°-175°, listed in Table 1. The power requirements of DBS satellites are so great that it is not practical to provide power in a single satellite for all the channels in a single slot. Accordingly, more than one satellite is assigned to each slot; and channels rather than satellites are assigned to individual applicants. Up to 32 channels per slot may be assigned. Current DBS channel assignments are listed in Chapter 2.

Canada Canadian satellites are located in the arc 104.5°-117.5° west longitude. Canadian satellites in orbit as of January 1990 are shown in Table 2.

Mexico and South America Mexico shares orbital slots with Canada. Because of their geographic separation, satellites can operate in the same frequency band and from the same slot without excessive interference.

South America shares orbital slots with the United States. The South American continent lies somewhat to the east of North America; and South American countries utilize an arc that partially overlaps the U.S., arc but extends to the east of it. Orbital slots assigned to Mexico and South America as of January 1990 are shown in Table 3.

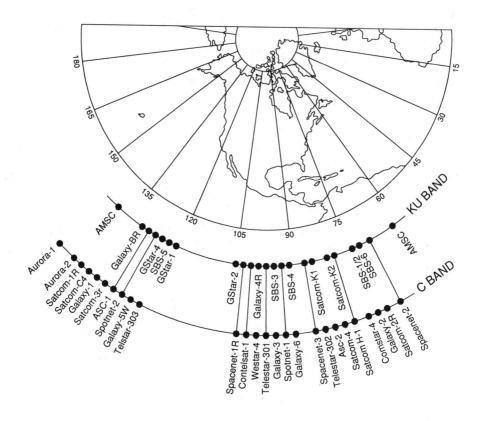

▶ **Figure 3** *United States orbital slot assignments.*

▶ **Table 2**   Canadian orbital slot assignments, January 1990

C-band Fixed Service Satellites (FSS)		
W. Long.	*Name*	*Operator*
104.5°	Anik-D1	Telsat
110.5°	Anik-D2	Telsat
Ku-band, Fixed Satellite Service (FSS)		
W. Long.	*Name*	*Operator*
107.5°	Anik C-1	Telsat
110°	Anik C-2	Telsat
117.5°	Anik C-3	Telsat

Source: *Broadcasting/Cable Yearbook.* Washington, D.C.: Broadcast Publications, Inc., 1990.

▶ **Table 3** Mexican and South American orbital slot assignments, January 1990

MEXICO

Hybrid C- and Ku-band Fixed Service Satellites (FSS)

W. Long.	Name
113.5°	Morelos-1
116.5°	Morelos-2
141.0°	Morelos-3
145.0°	Morelos-4

SOUTH AMERICA

Country or administration	Orbital slot	Satellite	Frequency
Argentina	80.0°	Nahuel-1	C/Ku-band
	85.0°	Nahuel-2	C/Ku-band
Brazil	61.0°	SBTS-B3	C-band
	61.0°	SBTS-C3	Ku-band
	65.0°	SATS-2	C-band
	65.0°	SBTS-A2	C-band
	65.0°	SBTS-B2	C-band
	65.0°	SBTS-C2	Ku-band
	70.0°	SATS-1	C-band
	70.0°	SBTS-A1	C-band
Colombia	75.0°	Colombia-2	C-band
	75.0°	Satcol-2	C-band
	75.4°	Colombia-1A	C-band
	75.4°	Satcol-1A	C-band
	75.4°	Satcol-2B	C-band
Cuba	83.0°	STSC-1	C-band
	97.0°	STSC-2	C-band
ASETA	72.0°	Simon Bolivar-C	C-band
(Bolivia, Ecuador	77.5°	Simon Bolivar-A	C-band
Peru, Venezuela)	89.0°	Simon Bolivar-B	C-band
	106.0°	Simon Bolivar-1	C-band

Source: *Broadcasting/Cable Yearbook*. Washington, D.C.: Broadcast Publications, Inc., 1990.

Satellite Look, or Elevation Angle

The *look* angle, the elevation of the path to the satellite above the horizontal, is critical to the performance of its transmission link. Three problems are encountered with low elevation angles that are just above the horizon:

The first is the difficulty in clearing buildings, trees, and other terrestrial objects. Failure of the path to do so may result in attenuation of the signal by absorption or in distortions from multipath reflections.

The second is atmospheric attenuation. The length of a low elevation path through the atmosphere before it emerges into space is much longer; and this increases rain attenuation, particularly when operating in the Ku band.

The third is the electrical noise generated by the earth's heat near its surface. This noise is picked up by the side lobes of receiving antennas aimed along a low-elevation path, and the signal-to-noise ratio of the link is adversely affected.

For downlinks, it is current design practice to utilize a minimum elevation angle of 5° for C-band satellites and 10° to 20° for K-band (see Chapter 4). For uplinks, the FCC requires a minimum elevation angle of 5° for all frequency bands, with possible exceptions for seaward paths where permission may be given for elevation angles as low as 3° (see Para. 25.505 of Part 25 of the FCC Rules).

The Prime Orbital Arcs

Because of the problems of transmission paths with low elevation angles, some orbital slots are more desirable than others. The orbital slot should be at a location which will result in a minimum elevation angle of 5° for C band and 10° to 20° for Ku band, for antennas located anywhere in the desired service area. The portions of the geosynchronous orbit in which the slots meet this condition are called *prime orbital arcs.*

The limits of the prime orbital arcs for 5° horizon clearance (C band) in the continental United States (CONUS) and selected points in Hawaii and Alaska are as follows:

	Degrees west longitude
CONUS	55-138
CONUS, plus Hawaii	80-138
CONUS, plus Anchorage	88-138
North Slope (Alaska)	120-190

The prime arcs for Ku band are shorter because of the need for greater elevation angles.

When earth stations are operated at northerly latitudes, the prime arcs are comparatively short, as shown in the previous table. Antenna elevation angles are low throughout Alaska, and satellites are below the horizon at the north pole.

Solar Eclipses

Solar eclipses occur when the earth comes between the sun and the satellite. They are not of direct interest to the users of satellite services, but they have a major effect on satellite design. The satellite depends on solar energy to provide electrical power to all of its systems, and this energy source is interrupted when the sun is eclipsed by the earth. Eclipses occur near midnight at the satellite's longitude for 21 days before and after each equinox, and their duration reaches a maximum of 70 minutes at the equinox. Batteries are installed on board the satellite to maintain a source of power during the eclipse period (see discussion later in this chapter).

Sun Outages

The elevated temperature of the sun causes it to transmit a high-level electrical noise signal to receiving systems whenever it passes behind the satellite and comes within the beams of the receiver antennas. The increase in noise is so severe that a signal outage usually results. The length and number of the outages depends on the latitude of the earth station and the diameter of the antenna. At an average latitude of

40° in the continental United States, and a 10-meter antenna, the outages occur for 6 days with a maximum duration of 8 minutes. With a less directional 3-meter antenna, the outages occur for 15 days, with a maximum duration of 24 minutes.

The times and durations of sun outages can be calculated precisely,[1] but this requires the use of astronomical tables that are frequently not available in the broadcasting industry. An easier way is to obtain the information from the satellite carriers, all of which have developed computer programs for outage calculations.

There is no way of avoiding sun outages; and if continuous service must be maintained, the only solution is to switch temporarily to another satellite.

SATELLITE LAUNCHING

Although users of satellite communication services are not directly concerned with satellite launching, it is useful for them to have a general knowledge of the process.

Placing the satellite in the geosynchronous orbit, 22,300 miles above the earth, requires an enormous amount of energy; and it probably would not have been possible were it not for the advances in rocketry that resulted from government military and scientific programs.

The launch process can be divided into two phases, namely, the *launch phase* and the *orbit injection phase*.

During the launch phase, the *launch vehicle* places the satellite in the *transfer orbit*— an elliptical orbit that at its highest point, the *apogee*, is at the geosynchronous elevation of 22,300 miles and at its lowest point, the *perigee*, is at an elevation usually not less than 100 miles. The energy required to raise the satellite from the elliptical transfer orbit to the circular geosynchronous orbit is supplied by the *apogee kick motor* (AKM), which is housed in the satellite.

There are two types of launch vehicles: *expendable rockets*, which are destroyed while completing their mission, and the *space shuttle* (more formally called the *space transportation system* or STS), which is reusable.

Expendable Rockets

Expendable rockets for communication satellites have three stages.

The first stage contains several hundred thousand pounds of a kerosene/liquid-oxygen mixture, plus a number of solid fuel rocket boosters that produce a tremendous display of flame—and ear-splitting noise—as the satellite lifts off its pad. It raises the satellite to an elevation of about 50 miles. The second stage raises the satellite to 100 miles, and the third stage places it in the transfer orbit.

After the satellite is placed in its transfer orbit, the rocket's mission is complete, and its remnants fall to earth. The satellite is placed in its final geosynchronous orbital slot by the AKM, which is fired on-command while the satellite is at the apogee of the transfer orbit.

The Space Shuttle

The Space Shuttle, or Space Transportation System (STS), performs the functions of the first two stages of an expendable launch vehicle. The satellite—together with the third stage, variously called an IUS (Integrated Upper Stage), SSUS (Spin

Stabilized Upper Stage), or a PAM (Payload Assist Module)—are mounted in the cargo bay of the shuttle. When the shuttle reaches its orbital elevation of 150 to 200 miles, the satellite and third stage are ejected from the cargo compartment. The third stage is fired, placing the satellite in the transfer orbit. The AKM in the satellite then places it in the geosynchronous orbit.

After all of its cargo has been jettisoned, the shuttle returns to earth for refurbishing and reuse.

History and Current Status of Launch Vehicles

All communication satellites launched from 1975 to 1982 used expendable rockets. In the early 1980s, the National Aeronautics and Space Administration (NASA) decided to abandon rockets and to move to the space shuttle. The space shuttle was designed to have the capability of putting very heavy satellites into orbit at a relatively low cost—the low cost resulting from the shuttle's reusability. Although development of the space shuttle was marked by serious technical difficulties, NASA was sufficiently confident of the shuttle's performance to cease taking reservations for expendable vehicle launches that would occur after 1984. Unfortunately, the technical problems were not solved on a timely basis; and they came to a climax with the tragic Challenger failure on January 28, 1986. At that time, only four commercial satellites (SBS, Telsat Canada, and GE Americom K1 and K2) had been shuttle-launched.

The defects in the space shuttle design, which were highlighted by the Challenger disaster, soon brought launches of commercial satellites by U.S. vehicles to a halt. Production of suitable expendable vehicles had virtually ceased, and lengthy engineering programs were required to solve the problems in the shuttle.

For a time, the Ariane rocket, a product of the European Space Agency (ESA), was the only available rocket for commercial satellites. ESA launched 7 U.S. satellites between 1984 and 1988, and 4 additional launches were scheduled for 1990. The Ariane, too, had its failures; and the McDonnell Douglas Delta series has the best record for commercial satellite launching.

In August 1986, a presidential directive was issued requiring the phaseout of the launch of commercial satellites by the government, and the reservation of the shuttle for scientific and military payloads. NASA and Air Force launch *facilities*, however, were made available to private companies; and contracts were negotiated with McDonnell Douglas (Delta rocket), General Dynamics (Atlas rocket), and Martin Marietta (Titan rocket) for the use of facilities in Florida and California. The first launch (an Italian satellite) under this program occurred in August 1989 with a Delta rocket.

COMMUNICATION SATELLITES

A communications satellite has two basic elements, the *space vehicle,* or *bus,* and the communications *payload.*

The Satellite Bus

As suggested by its name, the purpose of the bus is to provide a working environment for the payload, while maintaining it in its orbital slot. The bus includes the mechanical housing for the satellite, the systems required to provide primary power for its electrical and electronic components, and the *station keeping* facilities.

The Payload
The payload consists of the communication system components (see Chapter 3). They include the antennas, the receivers, and the transponders.

EARTH STATIONS

Earth Station Types
Earth stations vary widely in complexity and cost. At one extreme are the TVROs (television, receive only), that are used by individual homeowners. They are sometimes called *homesats* and may be purchased for less than $2000. The earth stations employed by major satellite carriers are at the other extreme. They may include (1) a number of very large antennas for communicating with several satellites simultaneously, (2) precision systems for tracking satellites, (3) uplink and downlink communication equipment, and (4) TT&C (telemetry, tracking, and control) systems for monitoring the performance of the satellites and maintaining them in the correct attitude in their orbital slots (station keeping). They are equipped with elaborate backup systems to provide a high degree of reliability, and their cost is measured in millions of dollars.

A description of the design and operation of the costly and complex earth stations employed by major satellite carriers is beyond the scope of this volume,[2] which is designed to give an overview of homesats, and the earth stations used by television broadcasting stations, major television networks, specialized and *ad hoc* television networks, cable TV systems, and television program suppliers. The equipment used in these earth stations is described in Chapter 4 and their system design is described in Chapter 5.

Uplink Earth Stations
A block diagram of a typical uplink earth station is shown in Figure 4.

If scrambling is employed to prevent unauthorized reception, the video and audio signals first pass through the *scrambler*. The scrambled signals are then fed to the *exciter*, the heart of the uplink transmitter. The exciter provides *preemphasis* and *sideband energy dispersal* of the baseband signals (see Chapter 4), generates and modulates an intermediate frequency main carrier and audio subcarrier(s), adds the audio subcarrier(s) to the video baseband signal, and "upconverts" the composite i-f signal to the final carrier frequency. The output of the exciter is fed to a *high-power amplifier (HPA)*, which is connected by waveguide to the antenna that radiates the signal to the satellite.

Downlink Earth Stations
A block diagram of a typical downlink earth station is shown in Figure 5.

The antenna receives the signal from the satellite and feeds it to a *low-noise amplifier (LNA)*, which provides the initial amplification. The output of the LNA is then fed to the receiver at the original carrier frequency through a waveguide. Alter-

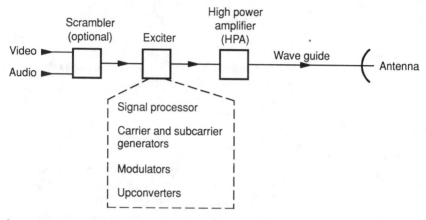

▶ *Figure 4 Uplink earth station block diagram.*

natively, a *low-noise block downconverter (LNB),* can be used, which translates all of the satellite's carrier frequencies of the same polarization to the first i-f frequency, usually 950 to 1450 MHz in the L band. This permits the connection from the antenna to the receiver to be made through a coaxial cable rather than a waveguide.

In an alternative configuration, only the LNA is mounted at the antenna; and its output, still at the original carrier frequency, is fed to the downconverter at the receiver by waveguide.

The antenna gain and the quality of the LNA are the key receiver components in determining its most basic performance specification, the signal-to-noise ratio. The quality of commercial LNA amplifiers varies widely in accordance with their cost. The difference in noise performance of a cyrogenically cooled LNA in a major earth station and an inexpensive homesat LNA may be 10 dB or more.

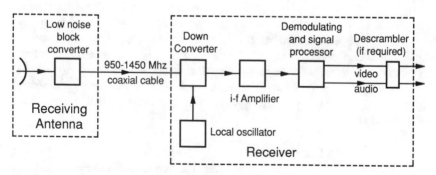

▶ *Figure 5 Downlink earth station block diagram.*

SUMMARY

A satellite communication system has two major components, the earth segment and the space segment. The earth segment includes (1) an uplink earth station that transmits a signal to the satellite and (2) a downlink earth station that receives and amplifies the signal transmitted back to earth by the satellite. The space segment is the satellite itself. It includes a number of transponders, each tuned to a single channel from the uplink. The transponder amplifies the signal and retransmits it to earth at a different frequency.

Launching a satellite into a geosynchronous orbit, 22,300 miles above the earth, is a major technical feat. The original launches of geosynchronous satellites employed expendable launch vehicles, that is, rockets. As a matter of government policy, the space shuttle was employed for a time; but a series of technical problems with the shuttle have caused the industry to revert to rockets for launching commercial satellites.

During the last two decades, the uses of satellites have expanded from intelligence gathering, military, and scientific, to include important commercial applications, including the transmission and distribution of radio and television programs, as described in Chapter 2.

Notes

1. See Loeffler, John, "Planning for Solar Outages," *Satellite Communications*, April 1983.
2. See bibliography for references.

2

Satellites in Radio and Television

THE UNIQUE ADVANTAGES OF SATELLITES

The importance of satellites to the radio and television industries has resulted from three unique characteristics: (1) The cost of satellite circuits is independent of their length—a signal can be transmitted by satellite across the country or across the ocean as cheaply as across the street. This makes satellite transmission cost competitive for some long-haul, point-to-point circuits within the United States; and, given the current limits on the bandwidth and capacity of undersea cables, it has been indispensable for transoceanic television circuits. (2) Downlink signals can be received over a wide area, a property which makes satellites a particularly cost effective and flexible medium for point-to-multipoint program distribution to cable TV systems, broadcast network affiliates, and to private homes. (3) Since uplink signals can originate over a wide area, satellites are useful or even indispensable for temporary communication circuits originating from mobile earth stations, particularly those which must be established on short notice, as in satellite news gathering (SNG).[1]

DEREGULATION

The usage of satellites for television and radio program transmission has been greatly increased by the policy of deregulation that the FCC has followed in recent years, particularly with respect to rates and earth station rules.

Rates

The first generation satellite communication companies—namely, AT&T, RCA, and Western Union— initially offered their services on a common carrier basis with all the restrictions of that mode. The FCC did not require common carrier operation, but it required carriers to file tariffs, and carriers were subject to rate regulation. Subsequently, the rates and other commercial policies of satellite carriers have become almost completely deregulated; and satellite users can now obtain services through a wide variety of sale and lease arrangements.

Receive-Only Earth Stations

The deregulation of receive-only earth stations in 1979 has had an even more striking effect on the usage of satellites.

FCC Rules originally required these stations to have a minimum antenna diameter of 10 meters, and to be licensed. To obtain a license, the applicant had to make a showing that the proposed site was *clear*, that is, that it did not receive objectionable interference from existing microwave systems or satellite uplinks. With these restrictions, the cost of typical receive-only earth stations used by cable TV systems was $75,000.

The requirements for minimum antenna diameter and licensing were eliminated in 1979. Most cable systems now use 5-meter antennas, and the size reduction—together with price competition and economies of scale—have reduced the cost of cable system earth stations to $5000 or less. Homesat earth stations with less demanding performance requirements can be obtained for less than $2000.

Obtaining a license is now optional for receive-only earth stations. A licensed receive-only earth station has the advantage of being protected against interference from future microwave and satellite systems.

Satellites and uplink earth stations continue to be subject to technical regulation (see Chapter 6). Its purpose is to prevent or minimize interference between satellites, between satellites and microwave systems, and between users of the same satellite. It is also intended to maximize the usage of the spectrum.

SATELLITE USAGE BY CABLE TV SYSTEMS

The distribution of programs to cable TV systems was the first major use of satellites by the television industry; and cable program suppliers continue to be the largest single market for satellite services. It is a synergistic relationship because satellite program distribution is responsible for the extraordinary growth of the cable TV industry.

History

In the pre-satellite era, the programming for cable TV systems was limited to off-the-air pickups from nearby broadcast stations and to programs imported by microwave from more distant stations. The variety of programs available to cable subscribers in the more populous metropolitan areas was not much greater than could be received off the air from local stations; and cable had little appeal in these areas. By 1977, the end of the pre-satellite era, the number of cable TV subscribers had grown to about 12 million, but the prospects for further growth were not encouraging.

Satellite distribution of cable TV program services began to grow rapidly in 1977; and cable systems were able to offer attractive programs which were not available from broadcast stations, even in the largest cities. With this incentive, the number of cable systems and the number of subscribers grew rapidly. In 1989, a little more than a decade later, the number of subscribers had quadrupled to more than 50 million.

This rapid growth was aided by the deregulation of the receive-only earth stations used by cable systems (see earlier discussion in this Chapter). Before deregulation, the $75,000 price tag for 10-meter earth stations was beyond the means of smaller cable systems. It was almost mandatory to group all program services on a single satellite so that they could be received with a single antenna. The licensing process was costly and time consuming; and it was sometimes impossible to get clearance for head-end sites where the earth station could be located—in accordance with the FCC's rigid interference standards.

With deregulation in 1979, these barriers were removed. Earth station prices came down rapidly, multiple-satellite transmission of different program services became practical, and the number of available services was no longer limited to the capacity of a single satellite. As of 1989, 122 transponders on 14 satellites carried cable TV programs. All but four of these transponders were on C-band satellites.

Satellite-Distributed Cable TV Program Services

Three types of program services, namely, basic, pay, and pay-per-view, are distributed to cable TV systems by satellite.

Basic services are provided by cable TV systems to their subscribers for a flat monthly fee. They include both off-the-air broadcast programs and programs distributed by satellite. Suppliers of basic services receive revenue by some combination of advertising and a small monthly per-subscriber charge to the cable system operator.

Pay or premium services are available to subscribers for the payment of an additional monthly fee, which is shared by the cable operator and the program supplier.

Pay-per-view services are available to subscribers for the payment of a fee for each program viewed.

Scrambling

The reduction in earth station prices had an unexpected effect: the unauthorized pirating of signals intended for cable TV systems by the use of *homesats*, that is, low-cost receive-only earth stations installed by members of the public. This led to the necessity of scrambling the signals of pay-TV services to prevent their reception except by authorized earth stations. (See material later in this chapter and in Chapter 4 for further discussion of scrambling systems.)

SATELLITE USAGE BY TELEVISION BROADCASTING

The television broadcasting industry adopted satellite program distribution more slowly than cable TV, but it has since become an integral part of the broadcasting medium. Most broadcast stations now have several earth stations that are used for the reception of network and syndicated programs and for the transmission and reception of news and sports programs from remote locations.

Public Broadcasting Service (PBS)

The Public Broadcasting Service (PBS) was the first major broadcasting organization to make extensive use of satellite program distribution. It began the transition from terrestrial circuits to satellites in 1978, using Western Union's Westar satellites. In the early years of this system, it was functionally equivalent to cable TV distribution systems—a one-way, point-to-multipoint service to PBS affiliates throughout the country. The experience was favorable, and it set a useful precedent for the adoption of satellite program distribution by the major commercial networks.

The Major Commercial Broadcast Networks

When satellite program distribution service first became available, the commercial television networks were under no particular pressure to adopt it. They had worked with AT&T and its terrestrial system for many years, and the relationship was generally satisfactory. The extremely high value and the perishable nature of their program services made the networks reluctant to risk service interruptions with an unproven system.

Nevertheless, there were problems with the terrestrial system that made a change desirable. AT&T's video circuits did not reach all the affiliates and could not provide service to or from points that were far removed from its switching centers. The networks were concerned that AT&T had the monopoly power to raise its rates to unreasonable levels. Finally, there was the positive example of PBS. As a result, the major networks began serious study of the use of satellites in 1980.

At present, commercial networks' operational requirements are more complex than those of PBS in its early years. PBS was initially a one-way distribution system, while commercial networks have complex two-way systems. Program segments often originate in several different locations, and multiple program schedules are distributed to different groups of stations. Their satellite distribution systems, therefore, require intricate switching facilities.

In spite of these concerns and problems, the advantages of satellites were a sufficient motivation for the networks to begin large-scale use of satellites in 1984 after years of study.

NBC   NBC was the first of the major commercial TV networks to utilize satellites for program distribution. NBC began the transition in January 1984 and completed it a year later. After long study, it opted to use the Ku band in CONUS (continental United States), concluding that its higher power and freedom from terrestrial interference outweighed the problem of occasional rain outages. It currently uses two Ku-band transponders on GE Americom's Satcom K2 satellite for its eastern and western feeds, and a C-band transponder of Satcom F1 for Alaska and Hawaii. In addition, it leases 13 transponders on Satcom K2 for feeds from affiliates and ENG.

CBS   CBS began the transition to satellites just 2 months after NBC, beginning in March 1984 and completing it in 1987. It chose C band, and currently leases four transponders—two on AT&T's Telstar 301 and two on Telstar 302.

CBS plans more extensive use of satellites in the future. It has announced the purchase of 12 C- and Ku-band transponders on Hughes Galaxy 4 and Galaxy 7, a hybrid satellite. The Ku-band transponders will be used for ENG uplinks. These satellites are scheduled to become operational in 1993.

ABC   ABC began its transition to satellites in late 1984 and completed it in 1987. It currently leases seven transponders on Telstars 301 and 302; and has announced the purchase of nine transponders, such as CBS on Galaxies 4 and 7.

Specialized and Ad Hoc Networks
The flexibility and relatively low cost of satellite distribution has led to the growth of specialized and *ad hoc* television networks. As with cable television, this development was accelerated by the deregulation of TVROs and their rapidly lower-

ing costs. Most commercial TV stations now have TVROs and are able to receive satellite signals.

Specialized networks offer programming for regional and national affiliates that is not available on the major networks. The distribution of television pickups of the games of a professional sports team to a group of stations on a scheduled basis is an example of a specialized network service.

Ad hoc networks are formed to distribute a single event to a group of stations.

Specialized and ad hoc networks frequently do not own and operate their own distribution networks, and they often find it more economical to lease it from *resale carriers*. Resale carriers lease facilities on a wholesale basis from satellite and terrestrial carriers, integrate them as required for a specific application, and rent them to the networks. The function of resale carriers is described in Chapter 8.

Program Syndication

Program syndicators are entrepreneurs who obtain the rights to programs and sell individual stations the rights to broadcast them. Their best customers are independent (unaffiliated) stations, but many network affiliates broadcast syndicated programs as well.

The original method of distributing syndicated programs was to record them on tape and "bicycle" the tape from station to station. Each station made a copy before forwarding it to the next station. While this was an acceptable method, it was not conducive to the highest technical quality.

As the number of stations equipped with TVROs increased, it became practical to distribute syndicated programs by satellite. Stations then receive and record the programs. As compared with bicycling, satellite distribution is faster, better, and, in some cases, cheaper. It is now an accepted industry practice.

The technical facilities required for the distribution of syndicated programs is the same as those for specialized networks, except that the program material is usually recorded rather than live.

Electronic News Gathering

In some respects the use of satellites for electronic news gathering (ENG), (or SNG for satellite news gathering), has had a more dramatic effect on the broadcasting industry than satellite distribution of network programs. Through the use of Ku-band earth stations, on-the-spot live coverage of news events has become practical in situations where C-band or microwave transmission are not possible. Most major television stations now have portable ENG earth station facilities for this purpose as well as to cover sporting events.

TV Broadcast Station Earth Station Facilities

A television broadcast station in a major metropolitan market typically has five or six earth stations. The complement might include primary and backup stations for receiving the network feed, a mobile SNG truck with Ku-band uplink and downlink facilities, a fixed Ku-band downlink station, and two or three C-band downlink stations for receiving syndicated programs and commercials.

DIRECT-TO-HOME BROADCASTING

Overview

The use of satellites for direct-to-home broadcasting was forecasted by industry pioneers in the earliest days of satellite communications. It has happened, but in a totally unexpected way.

It was originally believed that the power level permitted for C-band satellites was too low to be practical for broadcasting, and that direct-to-home service would require medium power Ku-band or high-power DBS satellites (see Chapter 3). To the surprise of the technical community, however, the public has been satisfied with the quality of C-band transmissions as received on homesats. The pirating by the public of C-band program services intended for cable TV systems was so widespread that it became necessary to scramble the signals so that they could be received only by authorized earth stations. On the other hand, primarily for economic reasons, direct-to-home service by medium-power satellites has been limited; and high-power DBS service has not become available in the United States.

The Direct-to-Home Broadcasting Market

Direct-to-home broadcasting must compete with cable; and, to date, it has been an unequal contest where cable is available. For a reasonable monthly fee, a cable TV system can deliver a multitude of satellite-distributed programs to the home, together with local and regional broadcast signals. When this service is available, households have very little incentive to install DBS receivers, unless the DBS programming is both unique and popular; and a supplier of unique and popular programming can reach a larger audience at a lower cost by distributing it through cable systems. Thus, the potential direct-to-home audience has been largely limited to households that have no immediate prospects for the availability of cable TV. In the United States this is estimated to be about 15 million homes. This is a sufficiently large audience to be a profitable increment for cable TV program services, but its economic viability has not been proven for services intended solely for the direct-to-home market—even for medium- or high-power satellites that can operate with very small and inexpensive receiving antennas.

The Power-Antenna Size Trade-Off

The design of direct-to-home broadcast systems is dominated by the trade-offs between satellite power, TVRO antenna size (and hence cost), and picture quality. For a level of picture quality judged to be satisfactory by the public, the trade-off for typical systems is as follows:

	Band		
	C	*Ku*	*DBS*
Transponder power (w)	4-10	45-60	150-250
Footprint coverage	CONUS	CONUS	1/2 CONUS
Antenna diameter (ft.)	4-10	2-5	0.75-2.0

A complete trade-off comparison must include the cost of the satellite. This favors C band because the cost of C-band transponders, whether leased or purchased, is considerably lower.

C-Band Program Distribution

The larger size of C-band antennas did not prove to be a serious barrier to the sale of C-band homesats. A trickle of installations began in 1980, and by 1982 it had become a flood. By 1987 the number of homesat installations was estimated to be nearly 2 million.

As the number of low-cost C-band homesats proliferated, signal pirating became a serious problem to the suppliers of pay TV service and cable TV operators. The solution was to scramble the signals so that special descrambling equipment, at first available only to cable TV systems, was required to receive a useable picture.

When scrambling began in 1987, there was a precipitous drop in the sale of new homesats. Sales recovered, at least partially, as descramblers were made available to the public; and at the end of 1989 it was estimated that more than 2.8 million homesats were in use and that they were being added at the rate of 30,000 per month.[2] General Instrument[3] has estimated that 10 million homes will be equipped with homesats by the mid-90s.

Ku-Band Program Distribution

The Ku band seemed to present an even better opportunity than C band for cable TV program suppliers to broaden their market to include homesat owners in areas not served by cable. Because of the higher downlink power density at Ku band, the required homesat antenna diameter would only be half as great as for C band, and this was expected to increase the homesat audience. As with C band, the same downlink signal could serve both cable systems and homesats, and distribution could be controlled by selective descrambling.

This concept has not been commercially successful to date. GE Americom and Home Box Office (HBO) formed a partnership in 1985, named Crimson Satellite Associates, to provide program services and to offer transmission service to other cable programmers on the GE Americom Satcom K1. The possibility of using smaller antennas in the Ku band was not a sufficient advantage to overcome the entrenched position and lower satellite costs of C band, and this venture was not successful.

A new proposal for the use of Ku-band distribution was offered in January 1990. Tele-Communications Inc. (TCI), the largest cable TV Multiple System Operation (MSO), announced an ambitious plan to organize a consortium of leading cable TV programmers that would provide direct-to-home service on 10 channels on K1. The consortium would include many of the most popular cable TV program services, and its future members are optimistic about its prospects.

Scrambling

Homesat manufacturers and the more than 2 million homesat owners mounted intense political opposition to the scrambling of satellite signals. It was even proposed that scrambling be forbidden by law, thus making signal piracy legal. The issue was resolved, in part by industry actions taken in response to government pressure, and in part by copyright legislation—the Satellite Home Viewers Act of 1988—which authorized all but broadcast network programs to be transmitted to home viewers.

The industry actions were directed toward making it possible for homesat owners to have access to scrambled signals at a reasonable cost. To accomplish this it was necessary to establish a technical center with the capability of authorizing descrambling on a selective basis and an administrative organization to receive orders, bill subscribers, and provide a list of subscribers to the technical center. There was a strong sentiment that the administrative organization should be a *third party* service, that is, that it be affiliated neither with program suppliers nor with cable TV operators.

General Instruments established the technical facility, the GI DBS Authorization Center in La Jolla, California, which transmits a code by satellite or telephone line to the transmission facilities of program suppliers. They then use the code to activate authorized descramblers.

The National Rural Telecommunications Co-op (NRTC) was the original administrative organization, and it has since been joined by a number of private companies that offer service packages.

With these facilities, it is now possible for homesat owners to receive scrambled satellite programs for approximately the same cost as cable TV subscribers.

Scrambling has not been totally successful in eliminating the piracy of satellite transmissions by unauthorized viewers. One set of industry figures[4] indicates that 1.1 million descramblers have been sold, but only about 500,000 viewers have subscribed to scrambled services. The presumption is that the other 600,000 have found ways to modify the descramblers to operate without authorization. The industry trade association, the *Satellite Broadcasting and Communications Association (SBCA)*, has formed the *Anti-Piracy Task Force (APTF)*, to investigate signal thefts by unauthorized modification of Videocipher equipment or other means.

Scrambling and descrambling equipment is described in Chapter 4.

DBS (Direct Broadcast by Satellites)

High-power DBS satellites, have not to date met the test of the marketplace in the United States for direct-to-home broadcasting. Their power is unnecessarily high for cable TV service, and the potential direct-to-home market (see above) has not been judged large enough to support the high cost of DBS transmission service, plus the cost of programming.

STC, a subsidiary of Comsat, ordered three DBS satellites in 1982—two to launch and one as a ground spare. After the satellites were completed, STC was unable to obtain partners or financing to launch the satellites and offer a program service, and the satellites were never launched.

In February 1990, a powerful consortium of NBC, Hughes Communications, Rupert Murdoch, and Cablevision announced plans for two or three high-power direct-broadcast satellites that would carry up to 108 channels. A key feature of this plan is a proposal to squeeze four television channels into each transponder through the use of digital transmission (see Chapter 1), a technology that has yet to be perfected.

As described in Chapter 1, applicants for DBS satellites are assigned channels, rather than complete orbital slots, because most satellites do not have sufficient capacity to provide primary power to all of the transponders in a single slot. Channels are usually assigned in pairs: one for eastern CONUS, and one for western CONUS plus Alaska and Hawaii.

The outstanding DBS channel assignments for the United States (January 1990) are listed in Table 4.

PRIVATE TELEVISION SYSTEMS

In addition to the major usage of satellites for the distribution of broadcast and cable television programs intended for the public, there is a small but growing application of satellites in corporate and other private television systems. Although either C band or Ku band can be used, Ku band is the most common because of its smaller antennas and the absence of potential interference from terrestrial microwave systems.

B-MAC Transmission

Since private television systems are closed-circuit, users are not required to transmit or utilize the standard NTSC format for color signals. B-MAC (Multiple Analog Components) is an attractive alternative. The signal information is sent in three components, *luminance*, Y, and *color differences*, R-Y and B-Y, rather than luminance and a color subcarrier(s) as in NTSC, PAL or SECAM. The subcarrier is at the upper end of the video spectrum where the noise is greatest in an FM system. Thus B-MAC makes it possible to obtain satisfactory results with a weaker downlink signal. The B-MAC format is also well adapted to scrambling.

A typical B-MAC earth station employs a 4.5 meter Ku-band antenna for both uplink and downlink.

▶ **Table 4**   DBS channel assignments (December 31, 1989)

Permittee	Authorized channels and orbital slot	
United States Satellite	8 ch @ 110°, 8 ch @ 148°	125 watts
Dominion Video	8 ch @ 119°, 8 ch @ TBD	200 watts
Hughes Communications	27 ch @ 101°, (Bandwidth compression)	180 watts
Advanced Communications	16 ch @ 110°, 16 ch @ 148°	125 watts
Direct Broadcasting Satellite	11 ch East 11 ch West	NA
Direct Satellite Corp.	11 ch East 11 ch West	100 watts
EchoStar	11 ch East 11 ch West	100 watts
Continental	11 ch East 11 ch West	200 watts
Tempo	11 ch East 11 ch West	100 watts

Source: *Broadcasting / Cable Yearbook.* Washington, D.C.: Broadcast Publications, Inc., 1990.

VSAT (Very Small Aperture Terminal) Networks

For teleconferencing applications with limited requirements for the portrayal of motion and resolution, satisfactory results can be obtained with VSAT networks that transmit digital signals at a 56 or 64 kbs rate.[5] The digital signal is processed for bandwidth compression, and picture quality adequate for teleconferencing can be achieved in this narrow bandwidth.

VSAT systems usually operate in the Ku band with 1.8 meter antennas. They are intended primarily for data communications, but extending their application to teleconferencing increases their value to their users.

INTERNATIONAL TELEVISION SERVICE

Satellites are used to a limited extent within the United States for long-haul, for example, coast-to-coast, television circuits; but this usage is not extensive because they must compete with terrestrial circuits. The most important application of point-to-point satellite television service is for transoceanic circuits where terrestrial facilities are limited or nonexistent. (This may change as fiber optic circuits become available.) Satellites are particularly useful for news broadcasts where real-time transmission is critically important.

Intelsat and Comsat

Initially, all intercontinental television traffic to and from the United States was carried on *Intelsat* satellites and *Comsat* earth stations in the United States and government-owned earth stations in other countries.

Intelsat is an international consortium, established by treaty between 108 countries, which owns and operates an extensive system of satellites for international communications services. Its ownership is shared by the members of the consortium in proportion to their usage of its facilities. Its primary focus is voice and data traffic, but it handles television signals as well. It owns no ground facilities, except for the TT&C stations (see Chapter 1) that control its satellites.

Comsat is a private company established by statute as the United States's chosen instrument to provide ground facilities for access to Intelsat satellites.

Originally, the services of at least four carriers were required to send or receive an overseas signal through the Intelsat/Comsat system: a U.S. carrier to connect the U.S. source or terminus with a Comsat earth station, a Comsat earth station to access the Intelsat satellite, an Intelsat satellite, and a foreign carrier (in most countries owned by the host government) to access the satellite at the other end of the circuit.

Competing International Services

Comsat and Intelsat were established as monopolies, but there was strong pressure in the United States to allow independent private companies to offer international television transmission service. Comsat and Intelsat opposed these proposals vigorously, and a debate developed reminiscent of the deregulation/divestiture issue with AT&T.

In 1984 the Reagan administration ruled that independent international television systems were required "in the national interest," and independent companies were permitted to apply for satellites to carry international television traffic. To protect Intelsat's economic interests, the ruling imposed strict conditions on these systems, the most important of which were a prohibition against providing a public telephone service and a requirement for "consultation" with Intelsat before a system could be authorized. As of this writing (1990), two independent satellite services have been authorized: PanAmSat to South America, and Orion Satellite Corp. to Europe.

In an equally important development, Comsat sold its earth stations to U.S. satellite carriers and now provides a booking and coordinating service for Intelsat service through these stations. In cooperation with U.S. carriers, Comsat established 18 international *gateways*, each a former Comsat earth station, or with a direct connection to one. The gateways provide the uplink/downlink interconnection with the Intelsat satellites.

Service through these gateways can be ordered from Comsat or from the carrier that provides the interconnection between the point of origination and the gateway facility.

In a further liberalization of its policies, Comsat now permits (but does not encourage) technically qualified earth stations owned by domestic carriers to access Intelsat satellites directly, thus bypassing the gateway facility. This service must be ordered and coordinated through Comsat by the domestic carrier.

SATELLITE USAGE BY RADIO

The impetus for the use of satellites by radio networks came from an increased demand for high-fidelity stereo transmission. AT&T's intercity audio circuits, although conditioned to provide more bandwidth than voice circuits, were not adequate for true high-fidelity service. Network stations faced increased high-fidelity competition from independent stations, which were programmed locally; and it was necessary for networks to match their sound quality.

The first radio networks to use satellites were the Mutual Broadcasting System, National Public Radio, AP Radio, and RKO Radio Network beginning in 1978 and 1979. These networks used analog transmission. NBC, CBS, and ABC did not begin the use of satellites until five years later in 1983 and 1984, using digital transmission to improve the quality further.

The digital system used for audio program transmission was first known as the Audio Digital Distribution Service (ADDS), and is now called the Digital Audio Transmission Service (DATS) (see Chapter 5).

Satellite program distribution has become standard for radio networks, both national and regional, and over 70 networks now use it.

TRANSPONDERS USED FOR TELEVISION PROGRAM TRANSMISSION

The listing of television services available on satellite transponders is volatile, and it should be updated frequently.[6] The number of transponders devoted to the major classes of television services at the end of 1989 was as follows:

	Number of transponders
Cable TV and Direct-to-Home	122
Broadcast - Full Time	20
Broadcast - Part Time and Occasional	146

The total usage of transponders by television program services, 288 in all, made the television industry the largest market for satellite transmission in the United States.

Notes

1. ENG, or electronic newsgathering, is generally used to designate any system that employs TV cameras (rather than photographic) at the scene. The TV signal may be transmitted to the station by microwave, satellite, or tape. SNG, or satellite newsgathering, designates an ENG system that employs satellites for transmission of the signal.
2. As reported by Satellite Broadcasting & Communications Association (SBCA), *Television Digest*, Vol 30, No. 2, January 8, 1990.
3. *Television Digest*, Vol 30, No. 5, January 29, 1990.
4. *Telecom Highlights International*, September 20, 1989, p. 18.
5. Inglis, A.F., *Electronic Communications Handbook*, Chapter 18, McGraw-Hill, New York, 1988.
6. The *Satellite Channel Chart*, published bimonthly by WESTAT Communications, PO Box 434, Pleasanton, CA 94566, lists the services carried on each transponder together with its polarization, the audio subcarrier frequencies, the audio subcarrier services, and the type of scrambling.

3

Communication
Satellites

SATELLITE CLASSIFICATIONS

Satellites are classified by their usage and their technical characteristics.

Usage Classifications

The FCC defines two basic usage classifications for United States communication satellites: fixed service satellites (FSS) and direct broadcast satellites (DBS). Fixed service is a broad general category that includes most communication satellites, while DBS is a special category for satellites that provide a direct television broadcast service to the general public.

Fixed Service is a generic term that is applied to communication services that are neither mobile nor broadcast. The application of this term to satellites is an extension of its application to earlier communication mediums, for example, microwave. Fixed service satellites are the principal subject of this book.

The DBS classification was established in response to a perceived large public demand for a satellite-to-home broadcast service. The DBS spectrum allocation and technical regulations were established in response to the economic necessity of using inexpensive earth stations with low sensitivity receivers and small low-gain antennas of limited directivity.

Technical Classifications

The most important technical classifications are frequency spectrum bands and service areas.

Frequency Spectrum Bands Three bands in the frequency spectrum are used for satellite communications: C band FSS, Ku band FSS, and DBS. The uplink and downlink frequencies allocated to these bands are as follows:

	Uplink	*Downlink*
C band, FSS	5.925-6.425 GHz	3.700-4.200 GHz
Ku band, FSS	14.0-14.5	11.7-12.2
DBS	17.3-17.8	12.2-12.7

The corresponding wavelengths are as follows:

	Uplink	Downlink
C band, FSS	5.06-4.67 cm	8.11-7.14 cm
Ku band, FSS	2.14-2.07	2.56-2.46
DBS	1.73-1.68	2.46-2.36

The characteristics of satellite services in these three frequency bands are quite different with respect to propagation effects, equipment performance, and FCC regulation. Accordingly, the band in which a satellite operates has a major effect on its performance.

Service Area   Satellite service areas vary widely in size, ranging from a metropolitan area to most of a hemisphere. The size of the service area is determined by the width of the satellite antenna beam, which, in turn, depends on the size of the satellite antenna—the larger the antenna, the narrower the beam. One classification system defines four area sizes, or beam widths: *global, regional, national,* and *spot.*

Global beams cover the entire area that is visible from the satellite. They are used by Intelsat for international communications.

Regional beams usually cover a group of countries, for example, western Europe.

National beams cover all or a significant portion of a single country. Most domestic U.S. satellites are in this category, although those that serve the continental United States as well as Alaska and Hawaii may have beam widths equal to those of regional systems.

Spot beams cover a limited area, and are used principally for point-to-point voice and data communications. They have found but limited application in television service.

Satellites designed for international voice and data communication often have a variety of antenna beam widths. The INTELSAT V satellite, for example, has two global beams, two regional beams, and two spot beams.

COMPARISON OF C-BAND AND KU-BAND SATELLITES

The most important differences between C-band and Ku-band FSS satellites for television applications are frequency sharing, antenna size, downlink power limitation, costs, and rainfall attenuation.

Frequency Sharing

C-band satellite uplinks and downlinks share frequency bands allocated to common carrier microwave systems. Ku-band satellite systems have exclusive use of their allocated frequency band.

Antenna Size

Ku-band antennas can be smaller because of the shorter wavelength.

Downlink Power Limitation

C-band satellites are limited to lower downlink power levels to avoid interference with terrestrial microwave systems. This also permits the use of smaller downlink antennas in the Ku band.

Earth Station Costs

In general, C-band earth station costs are lower, in part because they are presently manufactured in greater volume.

Satellite Costs

The cost of Ku-band transponders, either lease or sale, is substantially higher than C-band. Most Ku-band satellites have fewer transponders, and they are higher-powered. A typical monthly lease rate for a C-band transponder (preemptible) is $80,000, while for a Ku band it is $180,000. The cost differential is partially offset by the fact that it is practical to transmit two television signals with a single Ku-band transponder.

Rainfall Attenuation

C-band transmissions suffer little if any attenuation when passing through regions of high rainfall. Ku-band signal transmissions, by contrast, are subject to degradation from heavy rainfall, particularly in tropical and subtropical areas where cloudbursts occur frequently.

Current C- and Ku-Band Usage

As the result of its lower cost and freedom from rainfall attenuation, together with its advantage of being first in the marketplace, C-band service is more commonly used for television communications services at the present time. The Ku band, however, is almost universally used for portable SNG service because of its smaller antennas and freedom from interference to and from microwave systems. NBC uses the Ku band as the primary distribution medium for service to its affiliates.

COMMUNICATION SATELLITE DESIGN

A communication satellite includes two primary systems, the *bus* and the *payload*.

The Satellite Bus

The satellite bus has three major operating subsystems: 1) the power system, which provides a continuous source of primary power for both the bus and the payload; 2) a command and control system, which monitors the operating condition of the satellite and executes operational commands from the TT&C (tracking, telemetry, and command) earth station; and 3) a station-keeping system for maintaining the satellite in the proper attitude in its orbital slot.

Power System   Satellites operate all their electrical systems with solar power generated by arrays of *solar cells* that convert radiant energy from the sun into electrical energy. On-board storage batteries are used to provide a continuous source of power through the eclipse period. These are usually nickel-cadmium or nickel-hydro-gen to provide maximum life and reliability.

The power requirements of the satellites used for DBS are so great that it is not always practical to provide sufficient battery capacity to maintain power, even for the brief periods of the solar eclipses. The effect of signal interruptions in this service can be mitigated by locating the satellite to the west of the desired service area, thus causing the eclipses to occur after midnight.

Command and Control System   The command and control system includes instrumentation for monitoring all the vital operating parameters of the satellite, telemetry circuits for relaying this information to the TT&C earth station, a sub-system for receiving and interpreting commands sent to the satellite from the TT&C, and a command subsystem for controlling the operation of the satellite.

Satellite Housing Configuration   The configuration of the satellite housing is determined by the system employed to stabilize the attitude of the satellite in its orbital slot.

Three-axis-stabilized satellites employ internal gyroscopes rotating at 4,000 to 6,000 RPM. The housing is rectangular with external features, as shown in Figure 6. The solar cells are mounted on flat panels that rotate once each day about an axis parallel to the earth's axis so that they always face the sun.

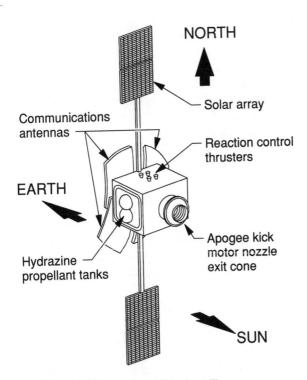

▶ *Figure 6 Three-axis-stabilized satellite.*

The nozzle of the *apogee kick motor* (AKM) is the exit cone for the final rocket stage that is housed in the satellite (see Figure 6). It has no further use once the satellite is in orbit.

An alternative design uses *spin stabilization.* The housing of a spin-stabilized satellite is cylindrical and rotates around its axis slightly more than once each second to provide the gyroscopic effect. The antenna must remain pointed in a fixed direction as the satellite spins; and it is *despun,* that is, connected to the body of the satellite by a rotating bearing.

In spin-stabilized satellites, the cells are mounted on the cylindrical surface of the satellite. Since only a fraction of the cells are facing the sun at any given time, a larger number of cells must be used.

Station Keeping   Although gravity and centrifugal forces are nominally in balance on the satellite, there are minor disturbing forces or perturbations that would cause the satellite to drift out of its slot if uncompensated. The gravitational effect of the sun and moon are examples. Another is the South American land mass which tends to pull satellites southward.

The physical mechanism for maintaining the satellite in its slot, *station keeping* is the controlled ejection of gas from thruster nozzles, which protrude through the satellite's case. The momentum of the gas stream produces an equal and opposite change in momentum of the satellite, and it is restored to its proper position.

Hydrazine gas is used for this purpose, and at the beginning of the satellite's life several hundred pounds of it are stored in the propellant tanks. It is emitted from the thrusters as required to maintain the satellite in its slot. The supply is eventually exhausted, typically after 10 years, and this usually ends the useful life of the satellite.

Beacons   Each satellite has two beacon transmitters—one for each polarization—operating at the ends of its 500 MHz spectrum. They transmit continuously and perform important roles in the operation of the system. They provide an aiming point for locating the satellite during the installation of a new earth station. They also provide the basis for automatic power level control for Ku-band uplinks. The strength of their signals is monitored; and if it is attenuated by rainfall, the uplink power can be increased to compensate.

The Payload

The communications satellite payload—its reason for being—has four major subsystems: the receiving antennas, the receivers, the transponders, and the transmitting antennas. They are combined in a variety of configurations in accordance with the satellite design objectives.

Frequency Reuse   The configuration of the payload is determined by the principle of frequency reuse, which is required for C-band satellites. Frequencies are used twice in the same satellite by cross-polarization of the radiated signals so that two transmission channels can occupy the same spectrum space. FSS communication satellites employ vertical and horizontal polarization on alternate channels, while DBS employs right- and left-handed circular polarization. See Figure 7 for the configuration of satellite channels with cross-polarization.

Payload Signal Path   The uplink signals pass from the satellite receiving antennas to the satellite receivers, each of which is tuned to a group of channels of a single polarization. The outputs of the receivers feed arrays of transponders, which shift the carrier frequencies downward to the downlink frequencies (hence the term *transponder*) and amplify them for transmission to earth. The transponders are the heart of the active satellite payload, and their power and bandwidth establish the most significant specifications of the satellite.

The number of transponders on a satellite can be limited either by the availability of spectrum space or of primary power. Spectrum space limits the capacity of most communication satellites, and their carriers completely occupy one of the frequency bands allocated for satellite communications. The availability of primary power may limit the number of channels per satellite for DBS.

Satellite Antennas   Satellite antennas are usually parabolic reflectors with feed horns located near their focal points. This format is very flexible, and many variations in design parameters are possible. Among these are the polarization of the radiated signal, the gain and directivity of the antenna, and the beam's shape and direction. The beam shape and antenna gain are determined primarily by the size and shape of the reflector—the larger the reflector, the narrower the beam and the higher the gain. The direction of the beam can be controlled within limits by the location of the feed horn with respect to the axis of the parabola.

Since satellites are usually designed to provide communication service to a relatively large area, the antenna beams must be broad, and the antenna gains comparatively low. The low antenna gain does not present a difficulty in the uplink where the satellite receives a strong signal from a high-power transmitter at the earth station. It is a problem for the downlink, particularly in point-to-multipoint service

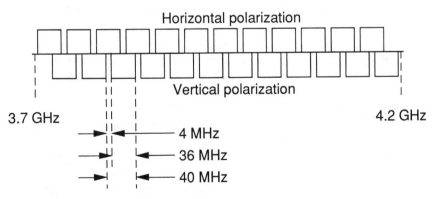

▶ *Figure 7 C-band channel configuration.*

where the cost of earth stations can be critical. Downlink design, therefore, often involves a number of significant trade-offs.

Satellite Receivers   Since communication satellites normally receive strong uplink signals, the performance of their receivers is less critical than that of the earth station receivers.

Satellite Transponders   The choice of transponder bandwidth and power are fundamental in the design of satellites.

For C-band satellites there is a *de facto* bandwidth standard of 40 MHz, including a 4 MHz guard band. Ku band does not have the same degree of standardization, but 54 MHz is commonly used.

The choice of power is more complex. Higher-power transponders are more costly, and this creates an economic trade-off between satellite and earth station costs. C-band downlink power is rather severely limited by FCC Rules. Finally, there are technical limits to the maximum power capability of satellite transponders.

Typical power ratings of satellite transponders currently in use (1990) are as follows:

Satellite type	Power (watts)
C band, FSS	8
Ku band, FSS	40
DBS	180 (proposal)

Early satellites used *travelling wave tubes* (TWTs) for the transponder power amplifiers. Developments in solid state technology have made it possible to use high-efficiency *solid-state power amplifiers* (SSPAs) in most FSS satellites. SSPAs are both more linear and more reliable than TWTs. TWTs are still used in high-power DBS satellites.

All transponder power amplifiers share a common characteristic known as *saturation*. This is the power level at which an increase in input power results in no further increase in the output power; in fact, it may decrease slightly. This condition, then, produces the maximum power output for the transponder.

For the satellite user, the most important measure of a satellite's power is its *effective isotropic radiated power (EIRP)*. This is the ratio (in decibels) between the strength of its downlink signal at a given point on the earth to the strength of a signal from an *isotropic radiator* that is radiating one watt from the orbital slot. (An isotropic radiator is an imaginary antenna that radiates equally in all directions. It cannot be physically constructed.)[1] Satellite operators publish EIRP contour maps of their satellites, and their use in system design is described in Chapter 5.

Transponder Traffic Capacity   The traffic capacity of a transponder is determined by its power and bandwidth.

C-band transponders are normally used for one television channel including video, audio, and possibly one or more narrow band communication or data channels. With their greater bandwidth and power, Ku-band transponders are frequently used to transmit two television channels (see Chapter 5).

C-BAND FSS SATELLITES

Channel Configuration

Each C-band communications satellite is allocated 500 MHz of spectrum space. Frequency reuse by cross-polarization is required by FCC rules; and it is universal practice to divide the satellite spectrum into 24 channels—each 36 MHz wide with 4 MHz guard bands, for a total of 40 MHz. This channel width follows a precedent set by microwave systems, which operate in the same region of the spectrum.

The channel configuration on a dual-polarized C-band communications satellite is shown in Figure 7. The center frequencies of alternate channels are spaced by one-half the channel width so that the unmodulated carrier and most of the energy of a modulated carrier falls in the guard band of the adjacent channels. Staggering the channel center frequencies, together with the isolation provided by cross-polarization, reduces the cross-talk between adjacent channels to a negligible level for television transmission systems.

Downlink Power Density

The downlink power density, that is, the radiated power per unit area and per kHz of bandwidth, is limited for C-band satellites by international agreement in order to minimize interference with terrestrial microwave systems.[2]

The downlink power density of a carrier that is frequency-modulated with a television signal and employing energy dispersal (see Chapter 4) is limited to an EIRP of about 38 dBw at low elevation angles to 48 dBw for elevation angles above 25°. The maximum EIRP of a typical C-band footprint is typically 34 to 38 dBW.

KU-BAND FSS SATELLITES

Channel Configuration

As with C-band, most Ku-band satellites employ cross-polarized feeds on alternate channels to double the spectrum capacity. The channel configuration is similar to that shown in Figure 7. Unlike C-band, however, there is no standard industry practice for channel bandwidth, and this varies from 36 to 72 MHz in different satellites. The lack of standardization makes it difficult to use cross-polarization as a technique for reducing interference between adjacent satellites (see Chapter 5).

Increasing the channel bandwidth increases the transponder power because the total satellite power is divided among a smaller number of transponders. It also makes possible a greater frequency deviation of the FM signal. The higher transponder power and greater frequency deviation both lead to a reduced requirement for the downlink antenna size. This advantage is offset to some extent, particularly in the high rain areas of the Gulf coast, by the need for a greater *fade margin* (see Chapter 5) for the Ku band.

A common configuration for Ku-band satellites is 16 cross-polarized channels, each with a 54 MHz bandwidth.

Downlink Power Density Limitations

Since Ku-band satellites do not share frequencies with terrestrial services, there is no FCC limitation on downlink power density.

HYBRID FSS SATELLITES
Description
Four hybrid satellites have been launched by GTE and Contel-ASC—three by GTE and one by Contel-ASC. The satellites are called hybrids because they have both C-band and Ku-band transponders—18 C-band and 6 Ku-band—for the four hybrid satellites launched by GTE and Contel-ASC. A total of nine hybrid satellites have been authorized by the FCC (see Table 1). They have a variety of configurations, but the FCC requires that new hybrids fully utilize the 500 MHz bandwidth of each band.

Applications
In addition to their greater capacity, hybrid satellites have the obvious advantage of providing the user a choice of frequency bands. They have the unique advantage that different bands can be used for the uplink and downlink portions of a signal path. This is accomplished in the satellite by a technique known as *cross-strapping*. Both CBS and ABC have announced future plans to use Ku band for ENG uplinks and C band for downlinks on hybrid satellites.

DBS SATELLITES
International System Specifications
DBS system specifications systems have been established by international agreement, most cently by a meeting of the World Administrative Radio Conference (WARC) in 1979; and, for the western hemisphere, by a meeting of the Regional Administrative Radio Conference (RARC) in 1983. The purpose of the specifications was to make it possible for members of the public to enjoy satisfactory reception with small receiving antennas (approximately 1 foot in diameter) and low-priced receivers. To achieve this objective the agreement contained the following provisions:

1. DBS was allocated a portion of the K band. This avoided the problem of interference with terrestrial microwave systems and made a higher downlink EIRP possible.

2. The maximum permitted EIRP was 56 dBW at the edge of the coverage area. It was recommended that the EIRP at the center of the coverage area be 3 dB higher than at the edge. This signal level would make it possible to achieve acceptable picture quality with an inexpensive receiver and an antenna as small as 1 foot in diameter.

3. Right and left-handed circular polarization would be used to achieve frequency reuse. Each polarization would have sixteen 24 or 27 MHz channels.

4. The spacing between orbital slots would be 9° rather than 2° as with the Ku band and C band. This is necessary because the effective beam width of a one-foot antenna in the DBS band is about 8°.

DBS Satellite Specifications
The specifications of three of the DBS satellites that have been built to date are shown in Table 5.

▶ **Table 5** DBS satellite specifications

Country	U.S.	Japan	Europe
Operator	STC*	NHK	
Number of transponders	3	3	2
Transponder bandwidth (MHz)	24	28	27
Frequency reuse	No	No	No
Polarization	Circular	Circular	Circular
Transponder power	215 W	200 W	230 W
Coverage	1/4 CONUS	Honshu	Sweden
EIRP (Edge of Coverage)	57 dBW	56 dBW	61.6 dBW (Spot)

* Built but not launched.
Source: *Broadcasting/Cable Yearbook*. Washington, D.C.: Broadcast Publications, Inc., 1990.

Notes

1. The use of an isotropic radiator as the reference antenna for satellite systems differs from the practice in television broadcasting where the half-wave dipole is commonly used. An isotropic radiator cannot be physically constructed, but theoretical calculations are more convenient.
2. The power limitation within any 4 kHz band is -152 dBw/m² for elevation angles below 5°, $[-152 + (\theta°-5°)/2]$ dBw/m² for elevation angles, $\theta°$, between 5° and 25° and -142 dBw/m² for elevation angles above 25°. See Appendix 3-1 for conversion to EIRP.

4

▼
▼
▼
▼
▼

Earth Station Equipment

This chapter describes the functions and performance specifications of the principal equipment components of the satellite earth stations used for the transmission and distribution of radio and television programs.

ANTENNAS

Antenna Types

Earth station antennas must be highly directional. In the transmit mode, they radiate energy in a narrow beam; and in the receive mode, they extract the radiant energy that arrives within the angular boundaries of this beam. The antenna's directivity is responsible for its gain, both transmitting and receiving, and makes it possible for adjacent satellites, separated by only 2° (9° for DBS satellites) on the orbital arc, to operate on the same frequencies.

Most satellite earth station antennas consist of a reflector and one or more feed horns, which illuminate the reflector with radiant energy in the transmit mode and collect it from the reflector in the receive mode.

Cross-polarized feed-horns with a separate feed for each polarization are used for antennas operating with transmitting or receiving channels of two polarizations.

Antennas with one feed-horn system have a single beam, while multiple feed horns can be used to produce multiple beams.

Since uplinks and downlinks operate on different frequencies, the same antenna can be used simultaneously for transmission and reception. This requires a waveguide filter in the coupling network to prevent leakage of transmitter power into the receiver.

Single-Beam Antennas In the most basic single-beam-antenna configuration, the reflector is a section of a paraboloid; and the feed horn, located at its focus, illuminates it directly. This is known as a *prime-focus-feed* antenna.

The dual-reflector antenna is an important variation. The feed horn is aimed away from the main reflector; its beam is intercepted by a subreflector and reflected back to the main reflector (see Figure 8).

Dual-reflector antennas come in a variety of configurations, depending on the shape of the two reflectors. If the subreflector is convex toward the main reflector as in Figure 8, it is known as a *Cassegrain* antenna. If it is concave, it is known as a *Gregorian*. In still another variation known as the *Gregux*, the subreflector consists of a concave ring.

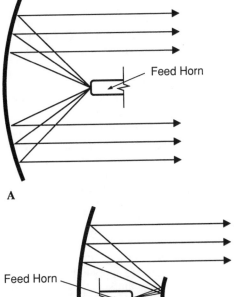

A

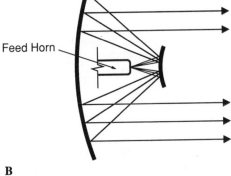

B

▶ *Figure 8* *(A) Prime-focus-fed antenna. (B) Dual-reflector antenna.*

Antenna designs are the result of a complex series of trade-offs between cost, directivity, and gain. Prime-focus-fed antennas are more inexpensive, suffer less loss of radiant energy by blockage, and, for lower gain antennas, have lower-level *side lobes*. Dual-reflector antennas have greater design flexibility and offer a greater choice of trade-offs. The two most common configurations used in satellite systems are the prime-focus-feed and Cassegrain. C-band receive-only stations frequently employ prime-focus feeds, while higher gain Cassegrain antennas are often used for C-band transmit-receive systems and for the Ku band.

Multiple-Beam Antennas If the feed horn is located on the axis of the main reflector, the antenna beam will be directed along this axis, but the beam can be pointed in other directions by offsetting the feed horn from the axis. This is known as an *offset-feed* antenna. Offset feeding degrades the performance of antennas somewhat, and there is little or no reason to use this technique in single-beam antennas. Its main application is in multiple-beam antennas.

Multiple beam antennas are used in the downlinks of earth stations that receive signals from several satellites, for example, at cable TV head-ends. They offer considerable economy, and occupy less space as compared with the use of a separate antenna for each satellite.

Multibeam antennas offer a considerable challenge to designers, and a variety of configurations has been developed. One uses a spherical rather than a paraboloidal reflector; another uses a combination spherical and paraboloidal surface and is called a *torus*.

In the simplest spherical configuration, feed horns are placed along the focal surface of the spherical reflector and directed toward its interior surface. The direction of each beam is coincident with the axis of its feed. This configuration has the advantage that antenna beams can be directed over a wide range of angles.

The shape of the torus antenna reflector is more complex. The surface is circular in the plane of the feeds, but parabolic in the orthogonal (right angle) plane. The use of a parabolic surface in one plane results in superior performance in some configurations as compared with the spherical antenna.

Electrical Performance Criteria

The most important electrical performance criteria of transmitting and receiving antennas are directivity, gain, and polarization isolation. They are reciprocal for the transmitting and receiving modes, and the same specifications apply to both. For receiving antennas, the *noise temperature* is also important, although in practice it is determined more by the elevation angle of the beam than by the antenna design. (See the definition of noise temperature later in this chapter and see Chapter 5 for its effect on the performance of a satellite system.)

Antenna Directivity   The radiation pattern of earth station antennas consists of a very narrow main beam surrounded by side lobes of much smaller amplitude (See Figure 9). The directivity of these antennas, that is, the extent to which they concentrate the radiated energy in a single direction, is important for two reasons: (1) It determines the effectiveness of the antenna in discriminating against signals from adjacent satellites when receiving, and in avoiding interference to these satellites when transmitting; and (2) It is the most important factor in determining the antenna's gain.

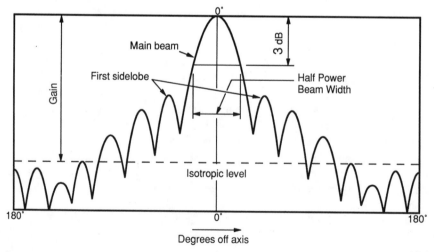

▶ *Figure 9 Earth station antenna radiation pattern.*

An antenna's directivity is completely specified by a radiation pattern, as shown in Figure 9. For many purposes, however, it is sufficient to specify two parameters, the *half-power beam width* (HPBW), and the *first side-lobe amplitude*.

The HPBW is the angle between the points on the radiation pattern at which the radiated power density is one-half its density at maximum. As a first approximation, the HPBW is proportional to the ratio of the wavelength to the diameter of the antenna reflector, or, for a noncircular reflector, to its dimension in the plane in which the beamwidth is specified. A trade-off can be made between beam width and side-lobe amplitude by controlling the uniformity of reflector illumination. Uniform reflector illumination gives the narrowest beam width, but also results in the largest side lobes—only 15 dB below the main lobe. A typical trade-off between beam width and side lobes occurs with edge illumination 13 dB below the center. The side lobe level is 23 dB below the main lobe for this design. By illuminating the center more strongly than the edges, the side-lobe amplitude is reduced, but at the expense of gain and HPBW. The trade-off between beam width and side-lobe amplitude is basic in antenna design.

Table 6 shows the HPBW and the beam width between nulls of typical C- and Ku-band antennas. The trade-off between HPBW and side-lobe amplitude in the antennas in these examples results in a first side-lobe that is 23 dB below the main beam. See Appendix 2 for the equations for calculating antenna beam widths.

FCC Directivity Specifications   The FCC has established specifications for the directivity of uplink antennas in order to minimize interference to adjacent satellites with 2° spacing. There are no mandatory requirements for downlink antennas, but the same standards are recommended. The protection to licensed downlink stations from co-channel stations is based on the assumption that these standards are met. Table 7 lists the maximum permissible gain envelope relative to an isotropic radiator in the angular range from 1° to 7° off the main beam and in the plane of the geosynchronous orbit.

Since the first null of most transmitting antennas is less than 1° from the axis of the main beam (see Table 6), it is the side lobes rather than the main beam which must be controlled to meet this specification.

Antenna Gain   The gain of a transmitting antenna is the ratio, usually expressed in decibels, between the power density radiated at the peak of its beam to the power density from an isotropic radiator (see Chapter 3).

The HPBW is the primary factor, determining the antenna gain—the narrower the beam width, the greater the concentration of radiated power and the higher the gain.

The gain is also determined by the antenna efficiency—the fraction of the power emitted by the feed horn that is radiated in the main beam. The antenna efficiency is determined by a variety of factors, such as power spillover at the edges of reflectors, power blocked by the feedhorn and subreflector, power absorbed by reflector surfaces, and power lost in the side lobes. The efficiency of practical satellite antenna designs varies from 50 to 85 percent.

The equation for calculating antenna gain is given in Appendix 3. Other factors being equal, it is approximately proportional to the ratio of the area of the reflector to

▶ **Table 6**   Typical antenna specifications

	Antenna diameter (meters)	HPBW (degrees)	Null BW (degrees)	Gain (dBi)[1]
C-band downlink (4 GHz, 7.5 cm wavelength)				
	2	2.6	6.4	36.6
	5	1.05	2.5	44.5
	7	0.75	1.8	47.4
	10	0.52	1.3	50.6
C-band uplink (6 GHz, 5.0 cm wavelength)				
	5	0.70	1.7	48.0
	7	0.50	1.2	50.9
	10	0.35	0.85	54.0
Ku-band downlink (12 GHz, 2.5 cm wavelength)				
	1	1.75	4.2	40.0
	2	0.88	2.1	46.0
	5	0.35	0.95	54.0
	7	0.25	0.60	56.9
Ku-band uplink (14 GHz, 2.14 cm wavelength)				
	2	0.75	1.8	47.4
	5	0.30	0.72	55.3
	7	0.21	0.51	58.4

[1] Decibels above the energy density from an isotropic radiator radiating the same power.

▶ **Table 7**   Maximum off-axis gain permitted by FCC rules (Plane of orbital arc)

Off-Beam angle (degrees)	Maximum gain (dBi)
1	29.0
2	21.5ts
3	17.0
4	14.0
5	11.6
6	9.6
7	8.0

Source: *FCC Rules and Regulations Satellite Communications*, Title 47, Part 25 of the Code of Federal Regulations. Washington, D.C.: Government Printing Office.

the square of the wavelength. The gains of representative satellite antennas, assuming efficiencies of 70 percent, are shown in Table 6.

Antenna performance is reciprocal for transmitting and receiving, but the gain is not the same in the receive mode as while transmitting because of the difference in uplink and downlink frequencies. Moreover, its effectiveness in extracting energy from a radio wave is determined not only by its gain but also by the wavelength of the radiation, since the received energy is proportional to the square of the wavelength. Thus a 2-meter antenna has a gain of 36.6 dBi at C band and 46.0 dBi at Ku band, but the higher Ku-band gain is offset by its shorter wavelength (one-third that of C band) and the received energy for the same EIRP is approximately the same. This suggests that antenna *area* rather than gain is the best indicator of its comparative effectiveness for receiving radiant energy at different wavelengths. Gain, however, is important for mathematical analysis and for comparisons at the same wavelength.

Polarization Isolation   Most satellites employ frequency reuse with horizontally and vertically polarized channels sharing the same frequencies (see Chapter 3). It is important that the antenna produce a high degree of isolation between cross-polarized channels, both for receiving and transmitting, to avoid mutual interference. See Chapter 6 for polarization isolation required by the FCC for uplinks. Satellite carriers typically require at least 30 dB for the uplinks to their satellites. This means that the strength of the cross-polarized components of uplink radiation must be at least 30 dB below the desired components.

Structural and Environmental Requirements

The antenna is the most exposed component of an earth station, and it is the most vulnerable to degraded performance or even destruction by adverse environmental conditions. These include wind, humidity, precipitation, icing, and the effects of salt atmosphere.

Mechanical Stability   A high degree of mechanical stability is a basic requirement of the antenna mounting structure. It must be capable of maintaining an acceptable pointing accuracy even in the presence of icing and high winds. Wind loading of the antenna can cause pointing errors; and with beams less than one degree wide, a small error can cause a serious loss of signal. Unfortunately, as the diameter of the antenna and its wind loading is increased, the amount of pointing error that can be tolerated is reduced. Larger antennas require an extreme degree of rigidity to minimize pointing errors in high winds.

Two specifications are normally stated for antenna stability, *pointing accuracy* and *survival*.

The pointing accuracy is defined in terms of the rms angular deflection (averaged over time) at stated levels of steady wind velocity and gusts. An example is 0.08° rms deflection with wind velocity of 45 mph gusting to 60 mph.

Survival is defined in terms of the wind velocity that can be tolerated from any direction without destruction or permanent damage to the antenna. An example is 100 mph without icing and 70 mph with 2 inches of radial ice.

The pointing accuracy of an antenna system is determined not only by the antenna structure but also by the stability of the base to which the antenna is mounted. The design of the base should take this into account, both with respect to short-term and long-term instabilities.

The specification for pointing accuracy is determined by the degree of reliability desired. Homesat antennas with relatively low requirement for reliability but with a requirement for low cost can tolerate a lower degree of pointing accuracy than antennas used in commercial service. Antennas used in uplinks must always have high pointing accuracy.

Other Environmental Factors   Humidity, especially in a salt atmosphere, can cause corrosion of electrical and mechanical components.

Precipitation brings with it the risk of leakage into sensitive electrical and electronic components.

Snow and ice forming on the antenna reflector can cause a serious degradation or even loss of signal. Deicers (see discussion later in this chapter) are needed to protect the antenna in climates subject to snow and icing if reliable performance is required.

Table 8 lists typical environmental specifications for commercial earth stations with high reliability requirements.

Antenna Accessories

Antenna Pointing Facilities   It is essential that the antenna mounting structure be provided with facilities for pointing the antenna toward the desired satellite. They may be pointed manually or by motorized remote control.

With manual pointing, the elevation and azimuth are adjusted independently, sometimes known as *elevation-over-azimuth.*

With motorized control, a *polar mount* is often used. The antenna is mounted so that it can be rotated around an axis which is parallel to the earth's axis. After the antenna is initially aligned so that its beam is pointed toward a slot on the geosyn-

▶ **Table 8**   Typical antenna environmental specifications

Parameter	Specification
Ambient temperature	-20° C to +55°
Wind loading (survival)	125 mph w.o. ice; 85 mph w. 2" ice
Pointing error (60 mph wind)	
C-band homesat;	
2-meter antenna	0.5°
Ku band;	
7-meter antenna	0.05°
Precipitation	1-in/h rain, or 0.25-in/h freezing rain, or 1-in/hour snowfall
Relative humidity	0 to 100 percent
Static ice load	0.25-in radial ice or 4-in snowfall

chronous orbit, it will follow the orbit quite closely over a wide range of azimuths as it is rotated around the polar axis. The visible satellites can then be accessed with no manual adjustment of the antenna.

Preset Positions   If the antenna must be moved from satellite to satellite frequently, a series of preset positions in the pointing mechanism is an extremely useful feature.

Polarization Control   The feed horn of the antenna must be adjusted so that it is aligned with the polarization of the satellite (see Chapter 5). This can be accomplished manually by rotating the feed horn at the antenna or remotely by a *ferrite polarizer*. The ferrite polarizer makes use of the *Faraday effect,* the rotation of the plane of polarization of an electromagnetic wave in the presence of a strong magnetic field.

Deicers   Deicing can be accomplished by means of electrical heating elements or by means of radomes. Electrical deicing may not be adequate in severe climates, and a radome around the reflector with a space heater to maintain its interior above freezing may be required.

SCRAMBLING SYSTEMS

The principal components of a scrambling system are 1) a data bank that stores the names of the authorized viewers and their address codes 2) equipment for scrambling the video and audio signals,[1] and 3) a data stream generator that transmits enabling signals to authorized descramblers.

Scrambling systems can be divided into two categories: hard and soft. General Instrument's *VideoCipher[R] I* and Leitch's *Digital* are hard systems. *VideoCipher II* and its successor, *VideoCipher II Plus*, are soft systems.

Hard systems are more secure, but their descramblers are more costly, and they are not economic for direct-to-home systems with millions of subscribers.

The VideoCipher II and its successor, VideoCipher II Plus, have become the de facto industry standards for cable programming.

The digital system manufactured by Leitch Video International, Inc., is used by NBC for its network satellite transmissions.

As described in Chapter 2, scrambled B-MAC (multiple analog component) transmissions are used in some private closed-circuit systems.

The VideoCipher II and II Plus Scrambling Systems

The VideoCipher II and II Plus scrambling systems employ relatively unsophisticated modifications of the video signal. It remains in analog form throughout the entire scrambling and descrambling process, and security is achieved by changing the synchronizing and signal format so that a standard receiver cannot reconstruct the image.

The scrambling of the audio signal is more secure than that of the video. It is converted to digital form and encoded with a digital encryption system (DES). The resulting digital pulse train is transmitted during the vertical blanking interval of the video signal. The descrambler includes a D/A converter.

As noted in Chapter 2, the security of the VideoCipher II system was frequently compromised, and a number of features were added to VideoCipher II Plus to make it more resistant to pirates. The encryption elements are all contained in a VLSI (very-large-scale-integrated) circuit that can be replaced with a revised version from time-to-time if necessary. The capacity of the system was also increased from 5 million to 50 million subscribers.

Descramblers are *enabled*, that is, authorized to operate on a channel, by an encoded digital control signal transmitted from the satellite to the earth stations. The control signal contains two levels of selective addressability, both of which must be received to activate the descrambler. The first is a continuous signal that is received by all the VideoCipher descramblers; the second is a periodic signal that provides authorizations for specific earth stations.

The control signal is generated by the GI DBS Authorization Center (see Chapter 2) and transmitted by satellite or telephone line to the uplinks used by the program suppliers. These uplinks, in turn, transmit the control signals to the subscriber earth stations. The DBS Authorization Center maintains a data bank of all authorized subscribers that it receives from the NRTC (see Chapter 2) or other authorized marketing organization.

VideoPal is a further development of VideoCipher that permits descrambling to be authorized on an individual program basis for pay-per-view systems.

The Leitch Digital Scrambling System
Broadcast network programs are not intended to be received directly from the satellite by the general public, and the potential number of authorized earth stations is much lower than for cable TV programs that are also received by homesats. This makes it economical to employ a hard scrambling system.

The Leitch scrambler first converts the incoming analog video signal to a digital format (A/D) and stores it in a memory. The digital signals in the memory are arranged in blocks of 120 lines, and the sequence of the lines as they are read out of the memory is randomly intermixed in accordance with a control data signal. The digital signal is then reconverted to analog form (D/A) for transmission over the satellite.

The inverse process occurs at the descrambler. The signal is passed through an A/D converter and stored in a memory; and the original line sequence is restored by the control data signal, which is transmitted simultaneously with the video. The control signal also provides the basis for selective addressability, that is, determining which earth stations are authorized to receive the program.

The security of the system depends on the security of the encryption of the control signal. This is enhanced by providing seven alternative super-encrypted keys installed in the descrambler, any one of which may be called up by the control signal.

As with VideoCipher, the audio signal is converted to digital form, encrypted by random alteration of time sequence, and transmitted as a QPSK (quaternary phase shift keying) subcarrier on the video signal.

UPLINK EARTH STATION EQUIPMENT

Figure 4 is a block diagram of an uplink earth station. In addition to the antenna, its principal components are a signal scrambler (optional), a signal processor and exciter, and a high-power amplifier (HPA).

Uplink earth stations may be either fixed or portable. Portable earth stations usually operate in the Ku band. Ku-band earth stations employ smaller antennas because of the shorter wavelength, and Ku-band satellite systems do not share frequency allocations with terrestrial microwave systems. Thus the earth station can be set up on short notice, for a major news event for example, without concern about interference with existing microwave systems.

The Signal Processor and Exciter

The signal processor and exciter (which may be combined or separate units) are the heart of the uplink earth station electronics; and they perform the functions of signal level control, preemphasis, sideband energy dispersal, carrier and audio subcarrier generation, modulation, and upconversion. Its output feeds the high-power amplifier (HPA), which provides the final amplification of the modulated carrier.

Video Preemphasis   It is an inherent characteristic of frequency modulation that the thermal noise in the output baseband spectrum increases linearly with frequency. This creates a special problem with NTSC (National Television Systems Committee) and PAL (phase alternating lines) waveforms because the color subcarrier frequency is in the high-noise region near the top of the baseband. In the absence of corrective measures, the signal-to-noise ratio of the chroma information in the picture would be unsatisfactory. This characteristic also creates a potential problem for the audio circuit(s) although possibly not as severe.

The standard corrective technique is to *preemphasize* the high-frequency components of the baseband signals at the uplink transmitter and to provide complementary *deemphasis* at the downlink receiver. Preemphasis is required by the FCC for FM broadcasting and has become a standard technique for satellite transmissions. It produces a major improvement in the signal-to-noise ratio of both audio and video signals.

Since the receiver deemphasis characteristic must be complementary to the preemphasis, an industry standard is required. A commonly followed standard for video is CCIR Recommendation 405-1. This specifies approximately 13 dB of preemphasis at 4 MHz as compared with low frequencies (see Table 9). This preemphasis characteristic improves the video signal-to-noise ratio by approximately 11 dB.

Audio Preemphasis   The audio preemphasis standard is the same as for FM broadcasting—the response of an L-R circuit in which the time constant, L/R, is 75 microseconds.

Sideband Energy Dispersal (C-Band)   Most of the energy in the spectrum of an r-f carrier frequency modulated by a video signal is concentrated in sidebands in the immediate vicinity of the carrier. This creates a problem for C-band satellite

▶ **Table 9** Video preemphasis characteristic–CCIR recommendation 405-1

Baseband frequency	Relative response (dB)
10 kHz	-10
20	-10
50	-9.5
100	-8.8
200	-6.8
500	-2.0
1 MHz	+1.4
2	+2.8
5	+3.6

systems because the FCC limits on downlink EIRP are based on the energy density per kHz of spectrum space.

The problem is overcome by adding a low-frequency energy dispersal signal to the video baseband signal before it modulates the carrier. This spreads the sideband energy over a larger region of the spectrum, reducing the energy density per kHz in the downlink signal and increasing the permissible EIRP.

In order to avoid deterioration of the baseband signal, the energy dispersal signal is synchronized with the television frame frequency—30 Hz for NTSC signals or 25 Hz for PAL. It is typically a sawtooth waveform having an amplitude that produces a 1 MHz peak-to-peak frequency deviation of the carrier. In a variation of this standard, the deviation is automatically increased to 2 MHz when no video signal is present to provide adequate energy dispersal under no-signal conditions.

Carrier Generation   In a typical exciter design, the carrier is generated at an i-f frequency of 70 MHz. The 70 MHz carrier is modulated with the video baseband and audio subcarriers and its frequency shifted upward to the final carrier frequency by means of an *upconverter*. The output of the upconverter feeds the *high-power amplifier*, which provides the final amplification of the signal for transmission.

Para. 25.202(e) of the FCC Rules requires that the frequency stability of the carrier generator be maintained within ± 0.001 percent.

The carrier frequency is typically adjustable in steps of 125 kHz for C band and 500 kHz for K band.

Audio Subcarriers   The program audio and auxiliary audio channels, for example, cue circuits, for television service are transmitted by means of frequency modulated subcarriers added to the video baseband. Exciters typically provide up to three subcarrier generators at frequencies that can be adjusted from 5.4 to 8.5 MHz. The most commonly used subcarrier frequencies for stereo transmission are 6.2 and

6.8 MHz, but this choice is not universal. See Chapter 5 for further discussion of the selection of subcarrier frequencies.

Subcarriers can also be used for the transmission of auxiliary audio services.

Modulator and Upconverter Filter   Under current operating practice, video signals are transmitted by satellite in an analog format by means of a frequency modulated carrier. The choice of the modulation index is an important design decision (see Chapter 5).

The modulated carrier is passed through a filter to eliminate any energy outside the limits established by the choice of modulation index. Exciters typically offer a choice of filter bandwidths ranging from 15 to 40 MHz. For C band, 36 MHz is a standard filter bandwidth. For Ku band, 40 MHz is a common bandwidth for single channel per transponder operation while 24 MHz is standard for dual channel (see Chapter 5).

Audio signals for radio service may be transmitted in either an analog or digital format. FM is the modulation mode for analog signals while some form of *PSK (phase shift keying)* is usually used for digital.

High-Power Amplifiers (HPAs)

The power level of the output of the exciter upconverter is typically of the order of one milliwatt, and the HPA must amplify this low-level signal to several hundred watts or even kilowatts to provide sufficient power to drive the satellite transponder.

A number of different power devices have been used for the final HPA amplifier stage, and the range of choices has broadened with the progress of technology.

The earliest options were klystrons and travelling wave tubes. Klystrons were capable of higher power, but their bandwidth was limited to a single channel. Travelling wave tubes (TWTAs) have bandwidths in excess of 500 MHz and can be used in systems in which more than one channel is amplified by a single HPA. Their power output was limited in the early years of satellite communications, but technical advances have increased their power capability significantly.

Advances in solid state technology have made solid state amplifiers (SSPAs) a strong contender in recent years.

It is a frequent practice to operate two HPAs in parallel at half-power and driven from a common source. This provides a degree of redundancy in the event of failure of one of the units and also prolongs their life.

Uplink Performance Specifications

The basic uplink performance specifications are the following: antenna directivity and gain, the HPA power, and the transmission parameters—which include bandwidth, frequency response, and linearity.

The antenna directivity must meet the FCC requirements as stated in Table 7.

Ideally, the uplink EIRP, the product of the antenna gain and HPA power, should be sufficient to drive the satellite transponders to saturation (see Chapter 3). The required EIRP depends on the characteristics of the satellite, but a typical value is 83 dBW. Achieving this EIRP from a single HPA may not be practical in a mobile truck because of constraints on antenna size and power availability (see Chapter 5).

The transmission parameters should be consistent with the EIA (Electronic Industries Association) video and audio system performance standards for satellite systems, EIA/TIA-250. See Tables 15 and 16 in Chapter 5. These standards are for the complete system, and the uplink alone should perform to tighter tolerances.

DOWNLINK EARTH STATION EQUIPMENT

Figure 5 is the block diagram of a typical downlink earth station. Its principal components, in addition to the antenna, are a low-noise amplifier (LNA) and downconverter, or low-noise block converter (LNB); an i-f amplifier and demodulator; a signal processor; and a descrambler (if required).

Downlink Video Signal-to-Noise Ratio

The video signal-to-noise ratio[2] is the most important single performance specification of a satellite circuit. It is largely determined by the specifications of the downlink earth station. Selecting the optimum trade-off between earth station cost and system signal-to-noise ratio is an essential step in the selection of downlink earth station equipment.

A satisfactory signal-to-noise ratio is more difficult to achieve with satellites than with microwave systems because earth station receivers must operate with much weaker signals from distant satellites. On the other hand, microwave systems are more subject to atmospheric fading. Thus the signal-to-noise ratio is the more important specification for a satellite circuit, while the fade margin is more important for microwave links.

G/T, the Figure of Merit

Other factors being equal, the signal-to-noise performance, S/N, of a downlink earth station is proportional to the ratio G/T, sometimes called the figure of merit, where G is the gain of the antenna and T is the *noise temperature* of the system, a quantity derived from the *noise figure*, F_n.

Noise Figure   The noise figure criterion was originally developed for specifying the performance of radar receivers. It is defined as the ratio of the signal-to-noise ratio at the output of a receiver to the signal-to-noise ratio at its input when the input is at room temperature, T_o. (T_o is assumed to be 290°K or 17°C):

$$\text{Noise figure} = N_f = (S/N)_{output}/(S/N_{To})_{input} \qquad (1)$$

The noise figure is a satisfactory criterion for systems in which the receiver is the only significant noise contributor, and it is still sometimes used to specify the performance of homesat receivers. Noise temperature has generally replaced the noise figure for calculating the performance of satellite systems because it simplifies the calculation of the combined noise contribution of all system components.

Noise Temperature   Most of the noise in a properly operating satellite communication system is *random*, with the same characteristics as *thermal noise*, the electrical or electromagnetic signals generated by a hot object. Although some of the noise in a satellite system is non-thermal in origin, it is useful to define an *equivalent noise temperature* that is applied to all system components. It is an expression of noise power (see Appendix 3 for the relationship between noise temperature and noise power), and is defined as the temperature at which a hot body would generate or emit thermal noise power equal to that generated or emitted by the component, whether thermal or non-thermal in origin.

The noise temperature, T, is specified in degrees Kelvin (°K), its elevation above absolute zero or -273 °C. Thus °K = °C + 273°. The noise temperature can indicate either the noise power at a point in the system or the ratio of the noise powers at the output and input of a system component. In either case it is proportional to the noise power, and a low-noise temperature is desirable. When applied to a component, the output noise power must be *referred* to the input, that is, adjusted for the power loss (as in a waveguide) or the power gain (as in an amplifier) of the component.

The relationship between the noise temperature, T, and the noise figure, F_n, is as follows:

$$T = T_o(F_n - 1) \qquad (2)$$

where $T_o = 290°K$ and F_n is expressed as an arithmetic (not logarithmic) ratio.

Because of its universality, the noise temperature is an important specification for most of the equipment components in a satellite downlink.

The use of the ratio, G/T, in calculating the system signal-to-noise ratio is described in Chapter 5.

Receiving Antenna Noise Temperature

The antenna system is one of the sources of electrical noise in a satellite receiving system, but the antenna itself is a minor source of noise. Most of the noise in the antenna output is thermal noise radiated by the earth and atmosphere. Satellite antennas can minimize atmospheric noise by their directivity and a sufficiently high elevation angle. This keeps the main beam well above the earth and shortens its path through the atmosphere (see Table 10).

It is usually undesirable to operate Ku-band antennas at elevation angles of less than 10° because of the relatively high-noise temperature of the earth and atmosphere and the high-signal attenuation during rain storms at low angles.

Earth Station Input Stages

Figure 5 shows a block diagram of the input stages of a downlink earth station. The LNA is combined with a downconverter to form a low-noise block down con-

▶ Table 10 Typical antenna noise temperatures

Elevation angle	C band	Ku band
5°	58°K	80°K
10°	50°K	58°K
30°	30°K	38°K

verter (LNB). The LNB amplifies the signal and shifts its frequency downward to the first i-f frequency, usually 950-1450 MHz. At this frequency, a coaxial cable rather than a waveguide can be used for the connection to the receiver.

The LNB is located near the antenna. This results in optimum performance because the signal is amplified before it suffers the attenuation loss of the waveguide or transmission line connecting the antenna to the receiver.

The noise temperature of an LNA or the LNA portion of an LNB is determined both by its design and the actual ambient temperature. Great progress has been made in the performance of LNAs in the past decade as the result of improvements in solid state components. *Field effect transistors* using gallium arsenide (GaAsFET) are commonly used in the LNAs of medium-priced earth stations, and a circuit known as a *parametric amplifier* can be used for higher performance. Still higher performance can be achieved by cooling the amplifier. Table 11 shows typical LNA noise figures and noise temperatures.

The LNA or LNB is perhaps the most critical element in determining the performance of a receiving system, and its cost\performance trade-off is an important decision in earth station design. The noise temperature of commercially available units varies widely, from 200° or more for low-cost homesats to 25° or less for the costly units used in major satellite carrier earth stations.

Receiver

Downconverter   The downconverter heterodynes the carrier-frequency output of the LNA down to an i-f frequency. It is the functional equivalent of the first detector or mixer in a broadcast receiver. A frequency synthesizer is often used as the local oscillator for tuning the receiver.

The downconverter can be located either in the receiver or it can be combined with the LNA at the antenna to form an LNB as shown in Figure 5. In the latter case, a second downconversion is usually employed to produce a lower intermediate frequency, for example, 70 or 230 MHz.

i-f Amplifier   The i-f bandwidth should be chosen to match the bandwidth occupied by the carrier and its sidebands. This, in turn, is determined by the modulation index of the uplink modulator and the audio subcarrier frequencies. Since the carrier-to-noise ratio is inversely proportional to the i-f bandwidth, it should be no wider than necessary to accommodate the carrier and its sidebands.

Some receivers have adjustable i-f bandwidths that can be optimized for different uplink modulation indices. Others have fixed bandwidths. For C-band receivers, typical values are 32 and 22 MHz. For Ku-band receivers, typical values are 40 MHz

▶ **Table 11**   Typical LNA noise figures and temperatures

LNA Type	C band		K band	
	F_n	T	F_n	T
Field effect transistor	1.1	85°	1.8	150°
Parametric amplifier	0.8	60°	1.5	120°
Cooled parametric amplifier	0.4	29°	0.6	43°

for single-channel per transponder transmissions and 20 MHz for dual-channel per transponder.

The term *effective noise bandwidth* is sometimes used to define the bandwidth of the i-f amplifier. This is the bandwidth of hypothetical amplifier with uniform response in its pass band and sharp cut-off at the band edges that has the same noise output as the amplifier being specified.

The gain of the i-f amplifier must be sufficient to saturate the limiter even with the lowest amplitude input signals. Further, its noise temperature should be low enough so that essentially all the noise in its output originates in its input.

Limiter, Demodulator, and Signal Processor   The purpose of the limiter is to provide a signal output that is free of residual amplitude modulation and is constant, in spite of wide variations in modulation and input level. The elimination of residual amplitude modulation may generate additional sidebands, and the bandwidth of the output circuit must be wide enough to accommodate these. If it is not, other amplitude modulated components will be regenerated.

The demodulator, or second detector, provides an output voltage which is proportional to the frequency of the input. Some demodulators have *threshold extenders* that permit the receiver to operate properly with relatively weak signals.

The signal processor removes the signal predistortions that were deliberately introduced in the uplink, such as frequency preemphasis and waveform spreading. It also filters extraneous r-f signals that may have passed through the demodulator.

Receiver Threshold

The signal-to-noise ratio of the demodulated video or audio signal, S/N, declines linearly with the signal-to-noise ratio of the carrier, C/N, until a critical point is reached, the *receiver threshold*. For values of C/N below threshold, S/N declines rapidly, and the signal becomes unusable.

The receiver threshold is a function of its design, and it is an important receiver specification. Typically it occurs at a C/N of 10 dB. Some demodulators have *threshold extenders* that lower the C/N threshold.

AUXILIARY EQUIPMENT

In addition to the basic earth station equipment described above, a wide variety of auxiliary equipment is available. They include test equipment, remote control and monitoring equipment, and automatic redundancy switches.

Test Equipment

An earth station should have available all the standard test equipment, such as signal generators and oscilloscopes, which are required for communication systems operating in the gigahertz frequency range. In addition, uplink stations should have three specialized types of equipment available: a spectrum analyzer, a signal level meter, and a loop test translator.

Spectrum Analyzer   A spectrum analyzer is perhaps the most useful type of test equipment for the initial setup and maintenance of an uplink. As its name implies, it displays the amplitude of the signals in a circuit as a function of frequency over a specified frequency range. It is not inexpensive — its cost ranges from $15,000 to $60,000, depending on its frequency range and other features — but a unit satisfactory for its specialized uses in uplink installation and maintenance can be obtained near the bottom of this range.

Among other applications, the spectrum analyzer can be used to identify a specific satellite during the installation of an earth station. It can be used to adjust the carrier frequency and modulation index of an uplink channel. Similarly, it can be used to adjust the frequency and modulation index of an audio subcarrier. It can be used to adjust the amplitude of an energy dispersal waveform. It is an extremely versatile unit.

Signal Level Meter   A signal level meter, calibrated for carrier-to-noise measurements, is required for routine checks of the carrier-to-noise ratio—a basic criterion of the performance of a satellite system.

Loop Test Translator   The loop test translator makes it possible to test the linearity and frequency response of a complete uplink/downlink earth station without passing the signal through the satellite. See Chapter 7 for a description of its operation.

Remote Control and Monitoring Equipment
Equipment is available for the remote control and monitoring of a variety of earth station functions.

Remote control equipment on the market covers a wide range of costs and the extent of control functions. A simple system provides facilities for the remote control of antenna azimuths and elevations. At the other extreme are computer driven systems provided with a screen that can display such information as block diagrams of the earth station subsystems, status and fault indicators for all components, and the elevation and azimuth of a number of satellites. They can also include automatic monitoring equipment such as a logging printer and alarms.

Automatic Redundancy Switches
Earth stations operating in systems that have high requirements for reliability usually have redundant, that is, spare, equipment components, which can be switched into use in the event of failure to the main component. Equipment is available which makes the switchover automatic.

If the system includes several channels, it is normal practice to provide only one redundant unit for all channels, on the assumption that it is very unlikely that two or more channels would fail simultaneously. If one redundant or backup unit is protecting N channels, it is known as 1:N protection. The redundancy switch equipment detects the failed channel and automatically switches the backup component into operation.

Special Equipment for Mobile Stations
Certain special equipment and facilities are required for mobile stations or trucks.

A good magnetic compass for orienting the truck with respect to magnetic north is an obvious requirement.

A truck stabilizer is necessary to establish and maintain the level of the antenna mounting structure within small tolerances.

It is essential to provide facilities for telephone communication with the satellite control center the signal receiving point, such as a broadcast station master control. This system can employ a combination of a cellular telephone and a two-way satellite circuit. The complexity of remote telephone systems sometimes exceeds that of the signal transmission system.

Notes

1. For a summary of scrambling techniques that provide security for video signals, see Inglis, A.F., *Electronic Communication Handbook*, Chapter 17, McGraw-Hill, New York, 1988.

   ® VideoCipher and VIDEOpal are registered trademarks of the General Instrument Corporation.
2. The term *noise* is used in communications systems to describe any unwanted electrical disturbance. It had its origin in telephony where these disturbances produced audible noise in the signal. Its use was continued in video transmission even though it produces no audible effect. In television, the visual effect of noise is *snow* in the picture.

5

Earth Station Planning and Design

This chapter describes the design practices that are applicable to the earth station types commonly used for the transmission and distribution of radio and television programming. They include the following:

1. TVROs for CATV head-ends and broadcast stations
2. Uplinks for CATV program suppliers
3. Uplinks for specialized and *ad hoc* broadcast networks, and for program syndication
4. Portable ENG (electronic news gathering) stations
5. TVROs for low-cost homesats

PERFORMANCE AND RELIABILITY SPECIFICATIONS

Establishing the desired specifications for system performance and availability is an essential first step in the design of satellite earth stations. It involves a careful consideration of the trade-offs between system cost, electrical performance, and reliability. This section describes commonly accepted industry standards for performance and availability that can serve as guidelines for earth station design.

Television Transmission System Performance Standards

The Electronic Industry Association (EIA) electrical performance standards for television transmission systems as stated in its publication, EIA/TIA-250-C[1], "Electrical Performance Standards for Television Transmission Systems," are commonly accepted industry standards for video and audio circuits.

EIA/TIA-250-C establishes performance specifications for six circuit lengths:

Short-haul	Less than 20 route miles
Medium-haul	20 to 150 route miles
Satellite	Independent of circuit length
Long-haul	Over 150 route miles
End-to-end	Complete circuit from source to end point

The video performance standards for the different circuit lengths reflect the relative difficulty in achieving them. For example, the specifications for satellite circuits are more stringent than for long-haul circuits because the latter are assumed to require a number of repeaters in tandem, each contributing additional degradation of signal quality. On the other hand, the signal-to-noise specification for satellite circuits (56 dB, weighted) is not as stringent as for short-haul terrestrial circuits (67 dB) or medium-haul circuits (60 dB).

The audio performance standards are the same for all circuit lengths with the exception of the signal-to-noise ratio, which is 2 dB less for end-to-end circuits.

The principal EIA/TIA-250-C transmission specifications for video and audio signals are summarized in Tables 12 and 13. This document also defines the parameters used to specify transmission performance and describes the procedures for measuring them.

▶ **Table 12**   EIA/TIA-250-C video performance standards

	Satellite	End-to-end
• Gain/frequency distortion		
200kHz (reference)	0.0 dB	0.0 dB
50 Hz to 150 kHz	±0.2 dB	±0.3 dB
At 3 MHz	±0.7 dB	±1.0 dB
3.3 to 3.9 MHz	±0.35 dB	±0.55 dB
At 4.2 MHz	±0.65 dB	±1.2 dB
• Chrominance-luminance gain inequality	±4 IRE	±7 IRE
• Chrominance-luminance delay inequality	±26 nsec	±60 nsec
• Field-time waveform distortion, p to p	3 IRE	3 IRE
• Line-time waveform distortion. p to p	1 IRE	2 IRE
• Short-time waveform distortion	2 percent	3 percent
• Insertion gain variation with time	±0.2 dB	±0.5 dB
• Luminance nonlinearity	6 percent	10 percent
• Differential gain 4 percent	10 percent	
• Differential phase1.5 degrees	3 degrees	
• Chrominance-luminance intermodulation	2 IRE	4 IRE
• Chrominance nonlinear gain distortion	2 IRE	5 IRE
• Chrominance nonlinear phase distortion	2 degrees	5 degrees
• Signal-to-noise ratio, unweighted	56 dB	54 dB

▶ **Table 13**   EIA/TIA-250-C satellite and end-to-end audio performance standards

• Amplitude-frequency response	
400 Hz (reference)	0 dB
At 50 Hz	-1.0 dB, +0.5 dB
100 to 7, 500 Hz	±0.5 dB
At 15,000 Hz	-1.5 dB, +0.5 dB
• Total harmonic distortion	0.5 percent
• Signal-to-noise ratio (unweighted)	58 dB[1]

[1] Specification for satellite link; end-to-end specification is 56 dB.

The video signal-to-noise ratio and system availability are the most critical of these specifications because they have a major effect on G/T and other factors affecting the cost of the downlink earth station. The performance of current systems, which often falls short of these standards, is described in more detail later in the chapter.

Note that the signal-to-noise ratio standard, S/N, is based on *weighted* noise. *Noise weighting* is used to estimate the perceived visual effect of the noise, and is based on the fact that the fine-grained noise generated by high frequency components is not as visible to the eye as the coarser noise generated by lower frequencies.

Noise weighting is accomplished by multiplying the noise frequency spectrum by a curve representing the relative response of a "normal" eye to noise components of different frequencies under a specified set of conditions. EIA/TIA-250-C specifies a standard weighting curve for 525 line NTSC systems, which is based on a viewing ratio (ratio of viewing distance to picture height) of 4.

The difference between the unweighted and weighted S/N expressed in decibels is known as the *noise weighting factor*. Weighted noise does not represent noise power but rather is an approximate measure of its visibility. The equivalent unweighted S/N, based on the actual noise power, is calculated by subtracting the noise weighting factor from the weighted S/N.

11 dB is the approximate noise weighting factor for NTSC FM systems, and this results in an unweighted S/N standard of 45 (56-11) dB for satellite circuits and 43 dB for end-to-end circuits that include a frequency modulated link. This is a conservative standard, and at this level the noise is virtually invisible.

If the viewing ratio is reduced, for example, to 3 times picture height, as has been proposed for HDTV, the noise weighting factor must be reduced correspondingly because fine-grained noise becomes more visible. The amount of reduction depends on the scanning standards of the HDTV system, but a 5 dB reduction would be typical. This would increase the unweighted S/N standard to 50 (45+5) dB.

FM Radio Transmission System Performance Standards

The audio standards in EIA/TIA-250-C are for the audio portion of a television signal. They would give excellent results for program distribution to FM radio broadcasting stations as well, and they are somewhat more stringent than the FCC standards. There is, however, a desire on the part of some broadcasters for even higher fidelity with stricter standards than those in EIA/TIA-250-C. These can be achieved by digital transmission. For example, harmonic distortion can be limited to 0.3 percent, and a signal-to-noise ratio of 60 dB or better can be achieved. The lack of noise in an idle digital channel is particularly impressive, and it can be as much as 80 dB below the reference program level.

Availability Specifications

As with the video signal-to-noise ratio, earth station costs are very sensitive to the availability specifications of the system. Maximum availability requires complete equipment redundancy and a large fade margin, both of which add greatly to the cost. The designer, therefore, must consider carefully the cost of occasional service interruptions and compare it with the cost of a more conservative design.

The need for a high level of availability varies greatly with the application. At one extreme is the transmission of a major event such as the Super-Bowl game, with millions of viewers and millions of dollars of revenue at risk with even a short interruption. On the other extreme is a homesat earth station where an occasional interruption is, at worst, an annoyance.

EIA\TIA-250-C specifies signal availability of 99.99 percent. This standard allows cumulative outages of 53 minutes per year from all causes—sun outages, equipment failures, operational errors, and, for Ku band, rain outages. This is a very demanding requirement, and the designer of the system should consider carefully whether its cost is justified.

EARTH STATION LOCATION

Site Requirements

There are three requirements for an earth station site as follows:

1. There must be a clear line-of-site path to the satellite.
2. The usual requirements for physical access, zoning and land-use restrictions, power availability, and proximity to the source or end point of the signal path must be met. (Under certain conditions the FCC has the authority to preempt state and local zoning regulations. (See Para. 25.104 of Part 25 of the FCC Rules.)
3. It must be free of objectionable mutual interference with other terrestrial systems, particularly microwave systems.

The first requirement, a clear line-of-sight path from the earth station site to the satellite or satellites to be accessed, is necessary for all frequency bands and earth station types. The second requirement, compliance with local land use and zoning ordinances, must be met for all stations. The third requirement, freedom from objectionable mutual interference, which is only of concern at C band, is quite different for licensed and unlicensed downlink earth stations. A licensed downlink station must meet FCC requirements with respect to co-channel interference that are almost as rigid as for uplink stations. The FCC definition of "objectionable" interference is quite severe, and many users are satisfied with the performance of a system that meets less rigid requirements.

Line-of-Sight Path   The availability of a line-of-sight path can be checked with a transit after the elevation and azimuth angles of the satellite have been determined. Although the initial requirement may be only to access a single satellite, conservative design practice dictates that there should be a line-of-sight to *all* orbital slots in the portions of the orbital arc allocated to the United States (see Chapter 1). Slot assignments may change, or there may be a need in the future to access a different satellite.

Tables may be available showing the angles to orbital slots from the earth station's vicinity, or they can be calculated from the equations in Appendix 4.

Site Clearance—Licensed Stations   The FCC requires that earth station sites for all uplink stations and for licensed downlink stations be *cleared*, that is, that they be coordinated with other users of the same frequency bands. For the Ku band, "other" users are Ku-band earth stations, and for the C band, they include both C-band microwave systems and other C-band earth stations. A new site must be cleared, that is, shown to be free of interference to or from existing and authorized facilities, before it can be licensed. FCC coordination requirements are stated in Paras. 25.201 to 25.204 of Part 25 of its Rules.

A preliminary examination of potential earth station sites can be made by field measurements to determine the existence of interfering signals. After a site has been tentatively selected on this basis, detailed coordination data must be prepared and filed with an application to the FCC (see Chapter 6). Applicants are urged to retain a professional frequency coordinator who is familiar with the FCC Rules and procedures, and who maintains a data base of terrestrial microwave facilities, existing and authorized.

Because of the large number of microwave installations, it is often difficult to find an earth station site in major metropolitan areas that can be completely cleared for C-band, that is, cleared for all channels and all orbital slots. For this reason, the major earth stations operated by satellite carriers are usually located many miles from the city center. Partial clearance, for example, for a single channel or orbital slot, can often be achieved, even in urban areas. Also, advantage can be taken of buildings or natural obstructions that are not listed in the FCC data banks.

Site Selection, Unlicensed Stations   The selection of a site for an unlicensed station is based on the field survey described earlier. The owner must make his own judgement as to whether the interference on any channels is severe enough to disqualify the site. Criteria for evaluating the severity of co-channel interference and techniques for minimizing it are described later as part of the treatment of downlink earth station design.

SATELLITE LINK CALCULATIONS

The design of earth station facilities requires calculation of the signal-to-noise ratio and *fade margin* of the satellite link. The fade margin is the difference between the carrier-to-noise ratio at the receiver under normal conditions and the receiver threshold (see Chapter 4). Calculation of the signal-to-noise ratio for frequency modulated systems is carried out in two steps, determining the *carrier*-to-noise ratio followed by the conversion of this ratio to *signal*-to-noise.

Carrier-to-Noise Ratio   The carrier-to-noise ratio is the ratio of the power of an unmodulated carrier to the noise power. It is given by the equation:

$$(C/N)_{dB}=[EIRP-20logf_{GHz}-10logB_{i-f(Hz)}+45] + (G/T)_{dB/K} \tag{3}$$

where:

EIRP = the effective isotropic radiated power of the satellite footprint at the downlink receiver location.

f_{GHz} = carrier frequency in GHz

$B_{i\text{-}f(Hz)}$ = i-f bandwidth *in Hz*

$(G/T)_{dB/K}$ = gain-to-temperature ratio as defined in Chapter 4. *G* is the gain of the downlink antenna and *T* is the noise temperature of the system, usually $T_{LNA} + T_{Antenna}$

Note that the carrier-to-noise ratio is inversely proportional to the i-f bandwidth. This creates the necessity for a trade-off between fade margin and signal-to-noise ratio in selecting the modulation index (see below).

Fade Margin

After the calculation of the carrier-to-noise ratio, C/N, the fade margin is easily determined by subtracting the receiver threshold from C/N. An adequate fade margin is particularly important in the Ku band because it should be large enough to offset rain fades.

Rain Fades

Unlike sun outages, rain outages are unpredictable. They are caused not only by the attenuation of the radio wave by water droplets but also by the presence of background sky noise which results from precipitation. Light-to-moderate rain rates do not cause a problem, and outages only occur in torrential rain storms. Uplinks usually have enough reserve power to overcome rain attenuation, and outages occur mainly in downlinks.

The cumulative outage time depends on the frequency of heavy rain storms, the earth station fade margin, and the elevation angle of the antenna. As noted earlier, elevation angles of less than 10° are unsatisfactory in the Ku band because of the long path of the beam through the atmosphere. It is fortunate that elevation angles are highest in the southern latitudes of the Gulf states where heavy rains are the most frequent.

Many attempts have been made to develop a method for forecasting availability on a quantitative basis by applying statistical measurement data to theoretical models. The results have not been completely satisfactory, and forecasts at best are order-of-magnitude estimates.

The required fade margin as calculated from these models for very high availability can be impractically large. One model, for example, indicates a fade margin requirement of nearly 40 dB for 99.99 percent availability in the Gulf Coast region. A practical solution in these areas is to make the G/T of the downlinks as high as economically practical and accept the availability that results. There is some evidence that the existing models are unduly pessimistic and that actual experience will be better.

On the basis of information currently available, the following fade margins are suggested for Ku-band circuits with a high availability requirement and with elevation angles in excess of 20°:

	Suggested fade margin
Gulf Coast states	16 dB
Other continental United States	13 dB

If an extremely high degree of availability is required, it can be achieved by the use of a C-band backup channel or by a *space diversity* Ku-band earth station installed some distance from the first. Torrential rain storms that cause fading usually cover only a limited area at any given time, and it is unusual to suffer simultaneous fading at sites separated by a few miles.

Signal-to-Noise Ratio

After calculating the carrier-to-noise ratio (C/N), it is converted to the signal-to-noise ratio (S/N) for video signals by means of equation 4 (all terms in dB):

$$(S/N) = (C/N) + F_{DEEMPH} + F_w + F_{FM} \tag{4}$$

where:

C/N is the carrier-to-noise ratio as calculated by equation 3.

F_{DEEMPH} is the deemphasis factor (see Chapter 4).

F_w is the noise weighting factor (see above).

F_{FM} is the FM improvement factor. It is the ratio of the S/N for the demodulated video baseband to the C/N of the unmodulated carrier. For satellite transmission of video signals, it is typically 20 to 30 dB. Its equation is:

$$F_{FM} = 20 \log[\Delta(f_v)/2B_v] + 10 \log[B_{i-f}/B_v] + 7 \tag{5}$$

where:

$\Delta(f_v)$ is the peak-to-peak carrier deviation

B_v is the video bandwidth

B_{i-f} is the i-f bandwidth

For NTSC transmissions, 12.8 dB is commonly used for the sum of F_{DEEMPH} and F_w.

Note that F_{FM} and, hence, the S/N increases as the square of the carrier deviation. Increasing the S/N by this method has the problem that it necessitates an increase in the i-f bandwidth and a reduction in the fade margin. As described earlier, the trade-off between fade margin and S/N must be carefully calculated.

UPLINK EARTH STATION DESIGN

The uplink earth station determines all of the properties of the downlink signal except for the EIRP, which is determined by the satellite. They include the type of modulation, the modulation index for frequency modulation, the bit rate for digital transmission, the r-f bandwidth, the use of subcarriers, and the baseband format, that is, analog or digital.

At the present time, the analog format and frequency modulation are used almost exclusively for satellite transmission of television signals, both video and audio, since the bandwidth requirements of digital transmission are excessive.[2] Both digital and analog transmission are used for radio signals.

▶ **Table 14**   Typical antenna diameter-HPA combinations

C band		
Antenna Diameter	*7 meters*	*10 meters*
Antenna Gain (dBi)	50.9	53.5
HPA Power (watts)	1,000	1,000
Max. EIRP (dBi)	80.9	83.5

Ku band			
Antenna Diameter	*3 meters*	*5 meters*	*7 meters*
Antenna Gain (dBi)	50.5	55.0	57.5
HPA Power (watts)	500	700	1000
Max EIRP (dBi)	77.5	83.5	87.5

The downlink earth station determines the signal-to-noise ratio and the fade margin of the link, within the parameters established by the uplink and the satellite.

Television Service
Antenna and HPA   The antenna directivity must be sufficient to meet the FCC Rules, as shown in Table 7; and its gain in combination with the HPA power should produce sufficient EIRP to drive a satellite transponder to saturation. The EIRP required to produce saturation is typically 83 dBW. The drive for multiple-carrier transponders is much lower, to avoid excessive intermodulation.

Table 14 is a tabulation of typical antenna/HPA combinations.

The HPA power in Ku-band uplinks can be adjusted, either manually or automatically, by measuring the strength of the downlink signal from the satellite beacon. If attenuation of the signal results from heavy rainfall, the HPA power can be increased to compensate.

Modulation Index   The choice of the *modulation index*, the ratio of the peak frequency deviation to the highest baseband frequency component, is a trade-off between the signal-to-noise ratio of the system and its fade margin. Increasing the deviation improves the signal-to-noise ratio of the receiver output, but requires a wider bandwidth i-f amplifier, which increases the r-f signal level required to exceed the receiver's threshold.

A typical C-band deviation standard is 21.5 MHz peak-to-peak, plus 1 MHz for the energy dispersal waveform. The i-f bandwidth is 32 MHz.

Although Ku-band transponders have a wider bandwidth than C-band, it is not always prudent to take advantage of this by a larger modulation index. To do so would require a wider i-f bandwidth resulting in a lower C/N and a reduced fade margin. An alternative is to utilize the wider Ku-band transponder bandwidth by transmitting two television channels on one transponder. In this mode, a typical peak-to-peak deviation is 18.2 MHz with an i-f bandwidth of 24 MHz.

Audio Subcarriers   Frequency modulated subcarriers added to the video baseband signal are used to transmit the audio program material and any auxiliary audio channels. Additional subcarriers are sometimes added to carry cue and other communication channels.

The choice of the subcarrier frequency is another trade-off. It must be far enough above the video baseband spectrum so that the two can be separated at the receiver by suitable filters. Also, increasing the subcarrier frequency permits a higher modulation index for the audio signal. The disadvantages are that it increases the bandwidth requirement of the system, reducing the carrier-to-noise ratio and fade margin; and it places the audio subcarrier in a higher noise region of the demodulated signal. The greater bandwidth reduces the carrier-to-noise ratio and the fade margin. Subcarrier frequencies commonly used are 6.2 and 6.8 MHz, although lower frequencies such as 5.4 MHz are sometimes used in dual-channel Ku-band systems to reduce the required i-f bandwidth.

There is no industry standard for the modulation index of the audio subcarrier, and it varies in current systems from the FM broadcast standard of 150 kHz peak-to-peak to as high as 900 kHz or even higher.

Mobile Uplinks   The extensive usage of satellite transmission for electronic news gathering and sporting events has created a demand for mobile uplinks or "trucks." Complete packaged uplinks (and downlinks) are now available from a number of manufacturers. These include all the facilities that are required for rapid setup and operation at remote sites that may not have a source of primary power. They are designed by the manufacturer so that the user has only the responsibility of selecting the one that best fills his needs, and, possibly, of specifying custom features.

The constraints on antenna size and HPA power that result from the mobility requirement may make it impossible to achieve a sufficiently high uplink EIRP to provide even one-half the power needed to saturate a transponder. The 3-meter, 500-watt earth station shown in Table 14 is near the top limit for the EIRP of mobile units. Commonly used are 2.4-meter, 300-watt units with an EIRP of 73 dBW.

Typical Transponder Utilization   For television service, the standard utilization of each 36-MHz satellite channel is one video signal, plus one or more audio signals. For less demanding applications, two television channels, including audio and video, can be transmitted in a single channel by reducing the modulation index. This mode is seldom used at C band, but it is common at Ku band.

Typical transponder utilizations for C-band and Ku-band transmission of television programs are summarized in Table 15.

Radio Service
Radio programs are transmitted in both the analog and digital modes.

Digital Transmission   PSK (phase shift keying) or FSK (frequency shift keying) is used for digital transmission. Several program channels can be multiplexed on a single carrier using time division multiplex (TDM), or a separate carrier can be used for each channel (single channel per carrier, or SCPC).

▶ **Table 15** Typical transponder utilization, television transmission

	C band	Ku band	
	One TV channel	One TV channel	Two TV channels
Transponder bandwidth	36 MHz	54 MHz	54 MHz
i-f bandwidth	32 MHz	32 MHz	24 MHz
P-P deviation, video	21.5 MHz	21.5 MHz	15.0 MHz
Audio subcarrier	6.8 MHz	6.8 MHz	5.4 MHz
P-P deviation, audio	900 kHz	280 kHz	150 kHz

The major radio networks employ a TDM digital system developed by Scientific Atlanta. This system multiplexes two 15 kHz stereo channels, one 7.5 kHz stereo channel, and two voice cuing channels in a 2.048 Mb/s bit stream. The transmission specifications of this system meet the requirements of high fidelity FM broadcasting stations.

The ability to multiplex cuing signals with the program audio is a feature of this system. This permits remote network control of station switching systems, cartridge recorders, and other functions.

Analog Transmission   Regional and specialized radio networks commonly use SCPC transmission with analog transmission and frequency modulation. The signal processing includes standard audio preemphasis and volume compression/expansion to increase the average modulation level. The frequency deviation is typically 150-200 kHz peak-to-peak.

DOWNLINK EARTH STATION DESIGN

Licensing
An FCC license is mandatory for uplink earth stations, but it is optional for downlinks.

Licensing has the advantage that the station is given protection against co-channel interference from future microwave and satellite systems. The licensing procedure, however, can be costly and time-consuming (see Chapter 6), and it imposes antenna specifications and site criteria that may be unnecessary.

The result is that very few homesats are licensed. On the other hand, cable TV systems and broadcast stations that depend on the station for their revenue may find it prudent to go through the licensing process for their TVROs.

Design Objectives
The principal design objectives for downlink earth stations are the following:
1. Cost.
2. Freedom from co-channel interference from terrestrial sources, such as microwave systems, and other earth stations, or from adjacent satellites.
3. A satisfactory signal-to-noise ratio.
4. An adequate fade margin, particularly for Ku-band stations.

The downlink design parameters that determine the performance against these objectives are the antenna location, the antenna directivity, and the earth station figure of merit, or G/T. The other key link parameters—the downlink EIRP, the modulation index, and the required i-f bandwidth—are established by the uplink and the satellite.

Cost

Downlink earth stations come in a wide range of costs, features, and performance. Simple homesat stations may cost less than $2000. TVROs for cable TV systems and broadcast TV stations cost from $5000 to $25,000. DATS TVROs for radio stations cost from $7000 to $9000. The cost of the downlink portion of earth stations used by major carriers may exceed $100,000. The cost variables include the antenna diameter, the receiver noise figure, the antenna mounting structure, accessories, operating features, the degree of redundancy, and the general quality of the system components. Establishing the cost/performance trade-off is a critical decision in earth station design.

Co-Channel Interference

The earth station licensing requirements as contained in Part 25 of the FCC Rules provide excellent assurance against objectionable co-channel interference. But to take advantage of the additional flexibility permitted in the location and design of unlicensed stations, it is necessary to consider the amount of interference that can be tolerated, the sources of interference, and the steps that can be taken to minimize it, even when it is not possible to meet the FCC licensing requirements.

Interference Visibility   As with FM broadcasting, the use of frequency modulation reduces the susceptibility of signals to co-channel interference. Interference begins to be visible at a carrier-to-interfering-carrier ratio (C/I) of 17 to 21 dB. It becomes objectionable, even to an untrained observer, at a C/I of about 9 dB. The FCC criteria for objectionable co-channel interference—28 dB for point-to-point service and 22 dB for point-to-multipoint service—are more stringent.

Terrestrial Interference Sources   The most common source of terrestrial co-channel interference in the C-band is microwave systems. They are a particular problem in congested urban areas that are termination points for multiple microwave paths.

The best protection against co-channel interference from microwave systems and other earth stations, of course, is to locate the station in an area that is totally free of interfering signals. If it is not practical to locate the earth station in an interference-free site, successful operation can often be achieved by making use of antenna directivity and the natural shielding provided by buildings or local terrain features. Alternatively, artificial barriers can be constructed. At C band, trees can provide 6 to 10 dB of isolation, earth berms 10 to 15 dB, and wire screens up to 20 dB.[3]

As a last resort, interference from microwave transmitters can be minimized by installing notch filters in the earth station i-f amplifier at the microwave carrier frequencies, which are spaced ± 10 MHz from the center of the transponder channel.

(Such filters are available commercially.) The elimination of the interfering signals is not complete, and the notches degrade the performance somewhat, but the results can be satisfactory for homesat systems.

Adjacent Satellites   The earth station must depend on the directivity of its receiving antenna and the suppression of its side lobes to protect it from interference from adjacent satellites. Antennas meeting the directivity standards established by the FCC for uplink antennas (see Table 7) will provide excellent discrimination against adjacent-satellite interference. In practice, receiving antennas with lesser directivity often gives satisfactory results, particularly for homesats with their less stringent performance requirements.

Signal-to-Noise Ratio Objectives

The EIA standards for signal-to-noise ratio shown in Table 12 are extremely demanding, and more moderate specifications have been developed by other industry and governmental organizations. Table 16 is a representative list of alternative standards. Some are stated in terms of unweighted noise and others in terms of weighted noise. Derived standards, assuming an 11 dB weighting factor for FM systems, are shown in parentheses. This assumes an NTSC system and a viewing ratio of 4.

The wide variation in these standards reflects the highly subjective nature of this aspect of picture quality and the great differences in the tolerance of members of the public to noise.

The weighting factor for HDTV signals is less, typically by 5 dB, than for NTSC systems. With this weighting factor, the HDTV unweighted S/N should be at least 5 dB greater for comparable visual perception.

Availability Objectives

The EIA availability standard of 99.99 percent is extremely difficult to meet, and, in some cases, may be impossible. Even 99.98 percent availability, which al-

▶ **Table 16**   Signal-to-noise ratio standards

Standardizing organization	Standard for	S/N (dB) weighted	S/N (dB) unweighted
EIA	Satellite link	56	(45)
EIA	End-to-end system	54	(43)
FCC	Cable TV system	(47)	36
TASO[1]	Broadcast TV		
Grade 1 (Excellent)		>(52)	>41
Grade 2 (Fine)		(44-52)	33-41
Grade 3 (Passable)		(39-44)	28-33
Grade 4 (Marginal)		(34-39)	23-28
Grade 5 (Inferior)		(28-34)	17-23
Grade 6 (Unusable)		<(28)	<17

[1] Television Allocations Study Organization

lows twice as much outage time, requires system redundancy and places severe demands on equipment reliability and earth station G/T. Such a system can cost twice as much as one with more modest requirements. For most applications, availabilities in the range of 99.5 to 99.8 percent are adequate and much more cost effective.

Earth Station G/T
The ratio G/T is the most basic specification of downlink earth stations because it determines both the signal-to-noise ratio and the fade margin (see equations 3 and 4). It is the ratio, expressed in decibels, of the antenna gain with respect to an isotropic radiator to the system noise temperature in degrees Kelvin (degrees centigrade + 273°). The result is expressed in dBK:

$$(G/T)_{dBK} = G_{dBi} - 10 \log T_K \tag{6}$$

PERFORMANCE OF REPRESENTATIVE SYSTEMS

This section describes the performance of five typical satellite systems used for the transmission or distribution of television signals. They are the following:
1. A high-performance system used for a point-to-point trunk circuit.
2. A system for the distribution of programs to cable TV systems.
3. A mobile Ku-band system for ENG service for a television broadcast station. (This system employs one-half transponder.)
4. A C-band homesat system.
5. A DBS system.

Uplink EIRP
Table 17 shows the uplink EIRP performance of representative antenna-HPA combinations. With the exception of the mobile uplink, in which the EIRP is limited by the constraints on antenna size and HPA power and by the sharing of the transponder with another channel, all have capacity in excess of the 83 dBW required to saturate the satellite transponder. They would be operated, therefore, at less than full power. The mobile uplink EIRP could be increased by 3 dB by operating two HPAs in parallel.

▶ **Table 17**   Representative uplink EIRP values

	Point-to-point trunk	CATV	ENG	Homesat	DBS
Band	C	C	Ku	C	DBS
Antenna					
Diameter, m	10	10	2.5	10	7
Gain$_{dBi}$	54	54	48	54	58
HPA Power (Max)					
Watts	2,000	2,000	500	2,000	2,000
dBW	33	33	27	33	33
Uplink EIRP$_{dBW}$(Max)	87	87	75	87	91

Downlink Earth Station G/T

Table 18 shows representative G/T values for downlink earth stations.

▶ **Table 18**   Representative G/T values downlink stations

	$(G/T)_{dB/K} = G_{dBi} - 10logT_K$				
	Point-to-point trunk	CATV	ENG	Homesat	DBS
Band	C	C	Ku	C	DBS
Antenna					
Diameter, m	11	4.5	5	2	0.5
Gain$_{dBi}$	55	42	54	36	34
Noise Temp. °K					
Antenna	40	40	50	40	50
LNA	50	85	150	125	150
T_K	90	125	200	165	200
$(G/T)_{dB/K}$	35	21	31	14	11

System C/N and Fade Margin

Table 19 shows representative carrier-to-noise and fade margin performance.

▶ **Table 19**   Representative C/N and fade margin values downlike earth stations

	$(C/N)_{dB} = [EIRP-20logf_{GHz}-10logB_{if.} + 45] + (G/T)_{dB/K}$ (5-1)				
	Fade margin = C/N - Receiver threshold				
	Point-to-point trunk	CATV	ENG	Homesat	DBS
Band	C	C	Ku	C	DBS
Downlink EIRP	38	38	43	38	50
f$_{GHz}$	4	4	12	4	12.5
B$_{if}$ x 10^6 Hz	32	32	24	24	27
G/T$_{dB/K}$	35	21	31	14	11
C/N$_{dB}$	31	17	24	11	12
Recvr Threshold$_{dB}$	12	8	12	8	8
Fade Margin$_{dB}$	19	9	12	3	4

FM Improvement Factor

Table 20 shows representative FM improvement factors.

System S/N

Table 21 shows the signal-to-noise performance (weighted) that is representative of these systems.

▶ **Table 20** Representative FM improvement factors downlink earth stations

$$F_{FM} = 20log[\Delta(f_v)/B_v] + 10log[B_{if}/B_v] + 7$$

	Point-to-point trunk	CATV	ENG	Homesat	DBS
Band	C	C	Ku	C	DBS
$\Delta(f_v)$ (MHz)	10.5	10.5	7.5	10.5	10.0
B_v (MHz)	4.2	4.2	4.2	4.2	4.2
B_{if} (MHz)	32	32	24	24	27
F_{FM}	23.8	23.8	21.5	23.8	22.7

▶ **Table 21** Representative signal-to-noise ratios (weighted)

$$S/N = C/N + F_{FM} + F_{DEEMPH} F_{FM}$$
$$F_{DEEMPH} + F_W = 13 \; dB$$

	Point-to-point trunk	CATV	ENG	Homesat	DBS
Band	C	C	Ku	C	DBS
C/N (dB)	31	17	24	11	14
Fade Margin (dB)	21	7	14	3[1]	6[1]
F_{FM} (dB)	24	24	21	24	23
S/N (dB)	62	48	52	42	44

[1]Includes use of threshold extender to reduce receiver threshold to 8 dB.

Homesat Performance

These tables clearly show the compromises which must be made to operate a C-band homesat with small antenna.

A weighted signal-to-noise ratio of 42 dB produces a picture that is noisy by professional standards, although it appears to be satisfactory to the public, particularly in the absence of competing program availability.

The fade margin of 3 dB is barely adequate, even when high reliability is not required. It is based on an assumed EIRP of 38 dBw at the station site, a threshold extender, and reduced i-f bandwidth. Reducing the bandwidth to 24 MHz to improve the fade margin will have an adverse effect on sound and picture quality.

If the EIRP were much less, it would be necessary to install a larger antenna.

The half-power beam width of 2.1° does not provide adequate discrimination against signals from adjacent satellites spaced at 2°. (At the present time, however, the spacing between most adjacent satellites is greater than 2°.)

All of these performance criteria could be improved, while reducing the antenna size by the use of Ku band. The present commercial impediments to Ku band are described in Chapter 2.

Notes

1. EIA/TIA-250-C Standard, "Electrical Performance Standards for Television Transmission Systems," was issued January 4, 1990 and superseded RS-250-B. It can be ordered from

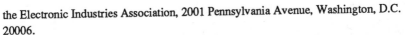

the Electronic Industries Association, 2001 Pennsylvania Avenue, Washington, D.C. 20006.

2. Compression techniques are being investigated that would reduce the bit rate and, hence, the bandwidth requirements for the transmission of video signals in a digital format. A target of 44 Mb/s has been established, which could be accommodated within a single C-band transponder. An even more radical proposal would squeeze four video channels into a single DBS transponder. The widespread use of these systems by the public is, at best, a number of years in the future.

3. Inglis, Andrew F., *Electronic Communications Handbook*, McGraw-Hill, New York, 1988.

6

▼
▼
▼
▼
▼

FCC Rules
and Procedures

OVERVIEW

Role and Authority of the FCC

The Federal Communications Commission (FCC) was established by the Communications Act of 1934, which gave it very broad authority to regulate interstate wired and wireless communications within the United States. Subsequently, the scope of its authority has been defined and broadened further by other statutes, among which was the Communications Satellite Act of 1962.

The regulation of satellite communications also requires international agreement. The International Telecommunications Union (ITU) governs radio services on an international level. It meets infrequently, and its year-to-year policy-making functions are handled by the World Administrative Radio Conference (WARC) and the Regional Administrative Radio Conference (RARC). To have the force of law, the decisions of these bodies must be specified in treaties that are ratified by the member countries.

Portions of the orbital arc have been allocated to the United States by treaty, and the FCC has the authority to assign orbital slots within these arc segments. Its policy for assigning slots was established by the *open skies* Order, which was issued in 1972. The policy states that any legally and financially qualified U.S. citizen or corporation is eligible to apply for a slot. It was an extremely important decision because, in contrast to previous practice, eligibility was not limited to existing common carriers. The financial resources required for the construction and launch of a satellite are large, greater than $100 million; and, in practice, this has limited satellite ownership to major corporations.

The procedures for the assignment of orbital slots is beyond the scope of this book, and this chapter is devoted to FCC Rules and procedures governing the installation and technical operation of uplink and downlink earth stations.

Deregulation

The open skies policy was the first major step in the FCC's steady progress in deregulating the satellite communications industry. Since then it has successively deregulated the nontechnical aspects of satellite communications, particularly the rates and conditions of lease or sale; and it removed the technical restrictions and mandatory licensing requirement for downlink receive-only earth stations. (The requirement for licensing uplink earth stations and the option of licensing downlink stations remain.) The salutary effect of deregulation on satellite service to the radio and television industries was described in Chapter 2.

FCC Rules and Regulations
The FCC's rules and regulations governing earth stations are contained in Part 25 of Title 47 of the Code of Federal Regulations. Parts 20 to 39 of Title 47 can be ordered (1990 price, $18.00) from the Superintendent of Public Documents, Government Printing Office, Washington, D.C. 20402-9371.

The rules concerning DBS stations are contained in the Final Acts of the 1983 Broadcast Satellite Service Regional Administrative Conference (RARC 1983). Among other provisions, this document specified the DBS orbital slots allocated to the United States and the maximum downlink power density.

TECHNICAL RULES

Uplink Earth Stations
The principal technical rules governing the operation of uplink earth stations have the purpose of minimizing interference to microwave receiving stations operating in the C band, 5925 to 6425 MHz. As a practical matter, no other services are authorized in the 14.0 to 14.5 GHz Ku uplink band.

There are four rules that address the interference problem: site requirements, Para. 25.203; power limits, Para. 25.204; minimum elevation angle, Para. 25.205; and antenna directivity, Para. 25.209.

Site Requirements   The FCC requires that a coordination procedure be carried out by the applicant for a new uplink earth station to demonstrate that harmful interference will not be caused to existing or authorized earth stations. The rules for coordination, described in Paras. 25.251 to 25.254, are complex and beyond the scope of this book, and applicants are advised to engage the services of a professional coordinator.

Certain preliminary actions can be taken before retaining a coordinator to screen out sites that clearly do not meet FCC requirements.

The first is to make field intensity measurements at the site over the complete frequency band—3700 to 4200 MHz for downlink stations and 5925 to 6425 MHz for uplinks. This will identify existing interference sources and also suggest the use of natural barriers for shielding.

The second is to consult with the Chief, Field Operations Bureau, Federal Communications Commission, Washington, D.C. 20554 if the proposed site is near an FCC monitoring station or a radio research facility operated by the Department of Defense or various civilian agencies.

The formal coordination process starts with the establishment of a *coordination distance*, which is defined as ". . .the distance within which there is a possibility of this earth station causing harmful interference to stations. . . sharing the same band. . . ." This is initially assumed to be 100 km (Para. 25.251(d)(2). The coordinator then applies various criteria to evaluate the possibility of interference.

Elevation Angle   Elevation angles of less than 5° for an uplink antenna are not ordinarily permitted because of potential interference to terrestrial circuits or possibly to adjacent satellites as the result of atmospheric scattering. Under special circumstances, the FCC may permit elevation angles between 3° and 5°. Under other circumstances it may require elevation angles greater than 5°. Aside from the FCC rule, it is usually bad practice in the Ku band to use an elevation angle less than 10° because of the long atmospheric path and severe rain attenuation.

Power Limits   The uplink power limit applies only to "bands shared coequally with terrestrial communication services," that is, to C band, and only at elevation angles less than 5°. It varies from an EIRP of 40 dBW in any 4 kHz frequency band at the horizon to 55 dBW at an elevation angle of 5°. Since elevation angles less than 5° are not permitted except in unusual circumstances, this rule seldom applies.

Antenna Directivity   The FCC requirements for antenna directivity in the arc from 1° to 7° from the axis of the main lobe and in the plane of the geosynchronous arc as seen from the earth station are shown in Table 7. Requirements for directivity in other planes and in the arc beyond 7° are described in Para. 25.209.
There is also an FCC requirement for off-axis cross-polarization isolation for C-band antennas. For off-axis angles, θ, from 1.8° to 7°, the required isolation is:

$$\text{Isolation} = [19-25\log\theta)]\text{dBi}$$

For values of θ from 7° to 9.2°, the required isolation is -2 dBi.
These values of isolation are in addition to those produced by the directivity of the antenna.
In addition, most satellite carriers have a requirement for on-axis polarization isolation, typically 30 dB.

Satellites
The FCC technical requirements for satellites that are of direct interest to users are the orbital spacing and the downlink power density limitation for C-band satellites.
Although some satellites are currently operating at spacings greater than 2°, the FCC's Rules and assignment policies are based on the assumption that all C-band and Ku-band satellites will be spaced by 2°. DBS orbital slots will be spaced by 9°.
The downlink power limitations apply only to C band and are described in Chapter 3.

Downlink Earth Stations
The original technical specifications for downlink earth stations were quite rigid. Since downlink stations do not cause interference, the primary purpose of the early specifications was to ensure high quality service. In the regulatory era, this was considered to be a legitimate role for the FCC. In the subsequent deregulatory era, it was considered more appropriate for the user to have the freedom to specify the service quality, particularly if the choice were between no service and service of

lesser quality. It was also considered desirable for the cost/quality trade-off to be made by the marketplace rather than by a regulatory agency. As a result, the FCC eliminated the requirement for licensing downlink earth stations in 1979. The effect of this decision on the growth of the satellite communications industry was described in Chapter 2.

There remains, however, the need for protecting earth stations carrying high priority or high value traffic from interference from new stations. To provide this protection, the owner of a downlink station is given the option of obtaining a license. In return, a licensed station is required to meet strict technical standards with respect to site location, elevation angle, and antenna directivity.

Site Location   The applicant for a licensed downlink station must follow the same coordination procedure and meet the same interference standards as an uplink station. For downlink stations, of course, potential interference will be from transmitting stations rather than to receiving stations. As with uplink stations, interference from terrestrial sources should not be a problem in the Ku satellite band since no transmitters are licensed from 11.7 to 12.2 GHz.

Antenna Elevation Angle and Directivity   Downlink antennas should meet the same standards for elevation angle and directivity as uplinks. The FCC will assume that these standards are being met in calculating interference from new stations.

APPLICATION FOR EARTH STATION LICENSE

Procedure
An application for a construction permit should first be prepared on FCC Form 401: "Application for Authority to Construct and Operate a Proposed Earth Station." This form includes the information necessary for the FCC to act on the application for license, Form 403: "Application for Radio Station License or Modification Thereof under Parts 23 or 25."

Requirements
The applicant must be legally and financially qualified, and the proposed station must meet the technical requirements of Part 25 as described above. The application must include a complete coordination study, which demonstrates that the proposed facility will not result in objectionable mutual interference between the proposed station and existing or authorized terrestrial stations.

INTERNATIONAL SERVICE

Three types of international television service are now available:

1. The program source or destination is connected to an international gateway (see Chapter 2) by a terrestrial circuit or a U.S. satellite carrier. An earth station at the gateway accesses an Intelsat satellite, and the program is transmitted to or from a

foreign *correspondent* facility by an Intelsat satellite. This service is ordered through Comsat, which makes all the arrangements and handles the scheduling.

2. Under certain conditions, U.S. earth stations other than those at the gateways can access Intelsat satellites directly. This saves the cost of the circuit to the gateway. This service is also ordered through Comsat.

3. After much controversy, two carriers were authorized to launch and operate satellites for international television service: PanAmSat to Latin America and Orion Satellite Corporation to Europe. Strict limits are placed on this service—most importantly that connection with public-switched telephone networks is forbidden, and that prior *consultation* with Intelsat is required to ensure that Intelsat will not suffer serious financial loss or *technical damage*, that is, excessive interference from the competing system.

7

▼ Earth Station Operation
▼ and Maintenance

Many of the routine operational and maintenance procedures for earth stations are similar to those employed in other complex electronic systems. This chapter describes those that are unique to satellite systems.

SATELLITE OPERATION

Locating the Satellite

The first requirement for placing an earth station into operation, either as an uplink or a downlink, is to locate the satellite. This involves five steps, (1) calculating the satellite's elevation and azimuth as seen from the earth station's location, (2) preliminary aiming of the antenna in accordance with these calculations, (3) final aiming of the antenna by searching for the satellite in the vicinity of the calculated elevation and azimuth, (4) adjusting the antenna's polarization, and (5) verifying that the correct satellite has been located.

Locating the satellite must be done with more care for an uplink than a downlink because a transmission to the wrong satellite or on the wrong polarization or channel can cause serious interference to other satellite users. *One must be absolutely certain that the antenna is pointed at the correct satellite, that it is adjusted to the correct polarization, and that the exciter is tuned to the correct channel before the HPA is turned on.* In addition, a communication link should be established to a control center maintained by the satellite operator or resale carrier to verify that the aiming and polarization of the antenna, the tuning and adjustments of the exciter, and the adjustment of the HPA power are correct.

Calculating the Satellite's Azimuth and Elevation   The satellite's azimuth and elevation angles can be calculated by means of the equations in Appendix 4 or by a calculator programmed for this purpose.

The variation in the angular location of the satellite within a limited area, say ±30 miles, is small—less than the error in aiming the antenna by compass and elevation indicator. A single set of calculated angles can be used by earth stations within this area, therefore, for the initial aiming of the antenna.

Preliminary Antenna Aiming

The antenna is first aimed in accordance with the calculated elevation and azimuth angles. The elevation is measured with an inclinometer and the azimuth with a compass. It is necessary, of course, to correct the compass reading with the local *variation*, the difference between true north and magnetic north. This can be obtained from U.S. Geological Survey maps or other sources. If the earth station is mobile, the truck is usually pointed to magnetic or true north by means of a compass and the azimuth of the antenna is measured with respect to the center line of the truck.

Final Antenna Aiming

Because of measurement inaccuracies, tropospheric bending, and the deviation of the earth from a perfectly spherical shape, it is unlikely that the initial aiming of the antenna on the basis of the calculated angles will be precisely accurate. For the final aiming of the antenna, the earth station is put in the receive mode and a small elevation-azimuth *box* around the calculated location is methodically scanned until a signal from the satellite is maximized. The receiver can be tuned either to a satellite beacon or to a transponder known to be operating on the satellite. This information can be obtained from the satellite operator or resale carrier.

Adjusting the Polarization

The polarization can be adjusted by tuning the receiver to an adjacent channel—which should be polarized at right angles—and rotating the antenna for *minimum* signal.

Verifying the Satellite

A spectrum analyzer is an invaluable tool in verifying that the correct satellite is being received. The 500 MHz spectrum of every satellite produces a characteristic *signature* on the analyzer that results from the types of traffic carried by its transponders. With the help of a listing of the traffic carried by each transponder, one can verify that the antenna is pointed at the desired satellite. With a little experience, it is possible to identify each satellite quickly.

UPLINK OPERATION

Exciter Adjustments

The uplink is the heart of the satellite transmission system because it determines the principle transmission parameters—the carrier frequency, the type of modulation, the modulation index for frequency-modulated systems, and the mode for transmitting the audio portion of the signal. The specific value of these parameters must be established by agreement between the satellite system licensee and the uplink and downlink operators.

Carrier Frequency Each transponder has a designated frequency band, and the carrier frequency and all of its sidebands must fall within this band. If the transponder is shared by more than one carrier, for example, a second video channel or digital data channels, the satellite licensee will specify the frequency band to be occupied by each as well as its share of the transponder power.

A spectrum analyzer enables the operator to set both the carrier frequency and the modulation index (see discussion later in this chapter).

Modulation Index   The trade-off between signal-to-noise ratio and fade margin that is involved in the selection of the modulation index is described in Chapter 5. Increasing the modulation index or peak-to-peak deviation improves the signal-to-noise ratio but reduces the fade margin. The modulation index is also limited in transponders carrying more than one signal because it is necessary to limit the spectrum space occupied by the sidebands. Some *de facto* standards have been developed by the industry, although their use is not universal as follows:

	C band	Ku band	
	1 Channel	*1 Channel*	*2 Channels*
Transponder bandwidth	40 MHz	54 MHz	54 MHz
i-f bandwidth	36 MHz	36 MHz	24 MHz
p-p video deviation	21.5 MHz	21.5 MHz	18.2 MHz

An alternative for a two-channel Ku-band system is a p-p video deviation of 15 MHz and an i-f bandwidth of 20 MHz.

The choice of the modulation index is a systems decision that should involve the customer, the uplink operator, the satellite operator, and the downlink operator.

The spectrum analyzer is used to measure the modulation index.

Energy Dispersal   An energy dispersal waveform (see Chapter 4) should be added to a video signal to be transmitted by a C-band satellite. This reduces the energy density per kHz in the downlink signal and increases the permissible downlink EIRP.

Again, the spectrum analyzer can be used to adjust the amount of frequency deviation introduced by the energy dispersal waveform.

Audio Subcarriers   The audio subcarrier frequencies and their modulation index should be set to the values established by the system design. Like the video modulation index, the choice of the audio subcarrier frequencies and their modulation index involves the customer, the uplink operator, the satellite operator, and the downlink operator.

The de facto standard for subcarrier frequencies are 6.2 and 6.8 MHz. A lower frequency such as 5.4 MHz is sometimes used to permit the use of a narrower bandwidth i-f amplifier and a greater fade margin.

The FM broadcast deviation standard of ±75 kHz with the broadcast preemphasis standard is often used with satellites, although other standards are permissible. The ubiquitous spectrum analyzer can be used to adjust the frequency and modulation index of the audio subcarriers.

HPA Power Adjustment

The uplink operator should maintain communication with the satellite operator or the resale carrier control center while the uplink power is being adjusted. For multi-channel transponders, the power should be set initially at a low level and adjusted upward so that it will not drive the transponder past saturation and adversely affect the performance of other channels.

DOWNLINK OPERATION

i-f Bandwidth

Most receivers have variable bandwidth i-f amplifiers. The bandwidth should be set to the narrowest bandwidth that accommodates all the significant sidebands in order to produce the largest fade margin. This can first be estimated by calculation and then determined empirically by reducing the bandwidth until the picture quality is significantly affected. (C-band systems operating with a large fade margin may use a somewhat wider bandwidth to produce the very best picture quality.) An approximate estimate of the bandwidth can be made from equation (7):

$$B_{i-f} = 0.75[\Delta(f_v)] + 2SC_{AUDIO} \quad (7)$$

where:

B_{i-f} is the i-f bandwidth
$\Delta(f_v)$ is the peak-to-peak deviation
SC_{AUDIO} is the audio subcarrier frequency

Interfering Signals

Each channel in the C band should be checked for interfering signals from terrestrial microwave systems. As noted in Chapter 5, it may be possible to reduce or eliminate the interference by erection of obstacles or by installing trap filters in the i-f amplifier at the location of the microwave carrier frequencies ±10 MHz from the center frequency of the satellite channel. This procedure should not be necessary, of course, at a site that has been cleared in accordance with FCC rules for licensed stations.

Measurement of the Carrier-to-Noise Ratio, C/N

The carrier-to-noise ratio is a critical criterion of the performance of the system, and it should be measured, both after installation of the link and periodically thereafter. It is measured in the receiver i-f amplifier. (This measurement is sometimes erroneously described as a signal-to-noise measurement. Measuring S/N is more difficult and yields very little additional information on the operation of the system.)

Measuring the C/N requires a signal level meter that has been designed for this purpose. The signal level of a carrier operating at full power is first measured. The carrier power is then removed (or the receiver tuned to a nearby channel without power), and the noise power is measured. The difference in power levels is the carrier-to-noise ratio.

The measured value of C/N should then be compared with the value calculated with equation (3) and the cause of any major discrepancy determined.

Sun Outages

Sun outages are only a problem to downlink stations, and downlink station operators should ascertain the time and duration of the outages as described in Chapter 1. The only solution for these outages is to utilize a backup satellite that is far enough removed in the orbital arc to prevent overlap of the outage times. For point-to-point service, the decision to utilize a backup satellite is an economic one. Is the cost of the service interruption sufficient to warrant the cost of the temporary lease of a backup satellite? For point-to-multipoint service, there is the additional problem that the outage times are different for all downlinks, and every downlink station would have to install a second antenna. For service to broadcast network affiliates or cable TV program subscribers, the downlink stations should be advised of the outage times and the steps, if any, that have been taken by the uplink operator to provide backup service.

COMMUNICATION SUBSYSTEM

It is essential that the operator of an uplink earth station have facilities for constant communication with a satellite control center while it is being set up. After the system is in operation, a communication link must be available at all times to the control center and to the TV station or cable system master control. This is particularly critical for mobile earth stations that are moved frequently and that may not have access to a public telephone. Cellular telephone systems are usually available, and most mobile trucks also have facilities for communicating via the satellite on a small carrier or subcarrier. While not technically sophisticated, these systems are often complex, and operating personnel should become thoroughly acquainted with them before taking to the field.

MAINTENANCE

This section describes the special maintenance procedures that are unique to satellite earth stations.

Signal-to-Noise Ratio

The most critical performance criterion for a satellite system is its signal-to-noise ratio, and a special maintenance program should be established to monitor this on a routine basis. The signal-to-noise ratio is directly related to the carrier-to-noise ratio; and since C/N is more easily measured than S/N, it is more frequently used as the criterion of system performance.

C/N, measured as described earlier, should be compared with the value calculated by equation (3), and any major discrepancies should be investigated. The most likely causes of a below-normal C/N are misalignment of the antenna or its components, or degradation of LNA performance; and the C/N measurement provides the most direct means of monitoring these system elements.

Linearity and Frequency Response

The linearity and frequency response of a complete earth station, including both an uplink and downlink, can be tested with a *loop test translator* without passing the signal through the satellite. The translator accepts a test signal at the output of the upconverter—in the range 5.925 to 6.425 GHz for C band or 14.0 to 14.5 GHZ for Ku band—and translates it to the downlink frequency, 3.7 to 4.2 GHz for C band, or to a standard first i-f frequency such as 950 to 1450 MHz. The signal then passes through the receiver downconverter, and the linearity and frequency response of the complete system are measured by standard techniques.

HPA Performance

The HPA is the uplink component that is most vulnerable to deterioration with age and usage, and the best criterion of its performance is the amount of drive required from the upconverter to produce the normal power output. If it becomes necessary to increase the HPA drive to produce the desired transponder drive—either to achieve saturation for a single channel or a predetermined power level for a multichannel transponder—it is a probable sign that the HPA is aging or that the antenna has become misaligned.

The antenna alignment should first be checked with the earth station in the receive mode, and if it is found to be correct, it is probable that the HPA is nearing the end of its life. If the station is carrying high-priority traffic, it would be advisable to replace the HPA. An even better solution is to employ two HPAs operated in parallel as described in Chapter 4.

SAFETY

General

It goes without saying that all of the safety standards published by the FCC, EIA, ANSI (American National Standards Institute), OSHA, and other local and state government agencies should be followed meticulously. This is in addition to following the dictates of common sense. For example, care should be exercised to keep the area around an antenna clear when it is being rotated to aim it at another orbital slot.

Radiation Density

Standards   The radiation density safety standard that is applicable to uplinks requires special mention. It is based on the potential hazard that results from the very high level of electromagnetic power density in front of a highly directional uplink antenna. Unfortunately, there is not complete agreement on the power density level that is considered safe.

Microwave radiation is *nonionizing*, that is, unlike X-rays, for example, it does not ionize molecules that are exposed to it. The original standards for nonionizing radiation were based on the assumption that the only injurious effects were caused by heating of the tissues. On this basis, a limit of 10 mW/cm² was established by OSHA in 1983 (Standard 1910.17).

Some medical specialists believe, however, that long-term exposure to nonionizing radiation has other harmful effects, such as damage to the eyes, or, possibly, that it is carcinogenic. ANSI, the industry standardizing group, adopted a more stringent standard of 5mW/cm² (ANSI C95.1) in 1982. The issue was confused further by the adoption of an extremely rigid standard of 5 microwatts/cm² by the Soviet Union, although the evidence for this standard was very questionable. The ANSI standard is generally followed in the United States today.

Safe Distances   Because of the highly directional properties of satellite antennas, it is not difficult to avoid exposure to unsafe levels of radiation provided that one stays away from the center of the beam.

The distance to the 5 mW/cm² energy contour for an antenna having a gain of 48 dBi, an EIRP of 83 dBW at the beam center, and complying with the FCC directivity requirements shown in Table 7 are shown in the table below:

Angle from beam center	Distance to 5 mW/cm² contour
0°	565 meters
>7°	<6 meters[1]

The conclusion is that radiation from an uplink antenna is not a hazard, provided that one stays clear of the beam center or the immediate vicinity of the antenna.

Notes

1. This calculation is only approximate because 6 meters is in the "near zone" of the antenna, and the pattern has not fully formed at this distance.

8

Satellite Services
and Earth Station Equipment

OVERVIEW

This chapter describes the satellite services and earth station equipment that are available in the marketplace. The businesses of supplying these services and equipments are dynamic, and directories of suppliers are subject to frequent change. For up-to-date listings, the reader is referred to standard reference yearbooks that are updated annually. Two of these are the following:

Broadcasting-Cable Yearbook
Broadcasting Publications Inc.
1705 DeSales Street, N.W.
Washington, D.C. 20036

and

Television and Cable Factbook
Services Volume
Warren Publishing, Inc.
2115 Ward Court, N.W.
Washington, D.C. 20037

SATELLITE SERVICES

Satellite services, that is, the right to use a satellite transponder, can be leased or purchased from a satellite *carrier* or from a *resale carrier*. Satellite carriers are licensed by the FCC to own and operate satellites and to offer satellite service for lease or sale. Resale carriers lease (or buy) earth and space segment facilities on a long-term basis and sublease them on a marked up, short-term basis. In effect it is a wholesale-retail relationship.

Satellite Carriers

There are presently (early 1990) four satellite carriers that sell or lease space segment facilities within the continental United States:

AT&T
Rm 4C103
Bedminster, NJ 07921

GE American Communications, Inc.
Four Research Way
Princeton, NJ 08540

GTE Spacenet Corp.
1700 Old Meadow Road
McLean, VA 22102

Hughes Communications Inc.
Box 92424
Los Angeles, CA 90009

Alascom Inc. provides interstate and intrastate satellite service in Alaska:

Alascom Inc.
Box 6607
Anchorage, Alaska 99502

International television transmission services can be ordered through the Communications Satellite Corporation (Comsat) (see Chapter 2):

Communications Satellite Corporation
World Systems Division
950 L'Enfant Plaza S.W.
Washington, D.C. 20024

Satellite Resale Carriers
Satellite resale carriers offer a number of useful services to the industry:
1. As described earlier, they fill the role of a wholesaler by leasing or buying satellite service on a long-term basis and subleasing it on a short-term basis to occasional users.
2. They act as systems integrators by providing network management for their customers and offering end-to-end service, employing a combination of satellite and terrestrial circuits.
3. In some cases they offer ancillary services such as recording program material for delayed transmission.

Site Coordination Services
The important role of site coordinators in obtaining earth station licenses was described in Chapter 6. Two site coordinators are the following:

Jefa International
Suite 280
1825 Plano Parkway
Plano, TX 75074

and

Comsearch
11720 Sunrise Valley Road
Reston, VA 22091

Terms and Conditions of Lease and Sale

The satellite communications industry is now unregulated, and both satellite and resale carriers have responded to this highly competitive environment by offering satellite services under a wide variety of terms and conditions. They include different *grades* of service, *periods* of service, and *scope* of service. This includes the choice of leasing or buying either a complete satellite, a complete transponder, or a partial transponder.

Grades of Service   Three grades of satellite services are offered—*protected*, *unprotected*, and *preemptible*. The supplier of protected service provides a backup satellite and/or transponder so that service can be restored quickly in the event of failure of the protected transponder. Unprotected service has no backup facilities, but it cannot be preempted. Preemptible transponders are the backups for protected service, and they can be preempted if necessary. Preemptible service is much cheaper, and it is particularly appropriate for occasional service that otherwise would be very costly.

Periods of Service   Satellite service can be leased for time periods ranging from an hour to several years. Many satellite users do not have a long-term need for full-time communication facilities, and it is common practice for them to lease service on an *occasional* basis, either from a satellite or resale carrier. It can either be a long-term commitment for part time service, for example, two hours per day, or truly occasional service, as for SNG applications (see Chapter 2).

Since capital costs—amortization and capital return—are the major cost elements for the satellite companies, the hourly rate for a long-term lease with a guaranteed revenue stream can be much lower than the occasional hourly rate. The resale carrier takes advantage of this by obtaining a lower hourly rate in return for a commitment to a long term lease. He takes the risk that he will obtain enough occasional or *part time* business to meet his costs plus a profit.

Scope of Services   The scope of service offerings is broad, ranging from a portion of a transponder to a complete transponder with uplink and downlink services. The usage of part of a transponder is, of course, subject to strict technical limitations to avoid degradation of other circuits on the transponder.

Teleports   Teleports are now available in most major cities that have earth station facilities and interconnecting terrestrial links. These are available for lease to occasional users as required to complete their satellite circuits.

Transponder Sales   As the result of deregulation, space segment service can be purchased as well as leased. A full transponder or a part of a transponder, for example, for an SCPC circuit, can be purchased. Title to the purchased facilities passes to the customer, who has the right to use them full time, subject only to the technical rules of the FCC and the satellite carrier. The satellite carrier continues to operate the satellite bus for a lump-sum payment or an annual fee. The satellite carrier may also provide backup transponders that would be available for use in the event of failure of the primary transponders.

EARTH STATION EQUIPMENT

Overview

The rapid growth in the use of satellites for communications has led to an equal proliferation in the number of suppliers of earth station equipment and the scope of their offerings. The result is that an overwhelming variety of earth station equipment is now available in the marketplace. A systematic approach should be followed in selecting the items of equipment from this abundance that best meet the requirements of the application. Chapter 5 describes the basic factors involved in designing a system and selecting its equipment complement.

Consulting Services

An earth station customer may find it desirable to obtain consulting advice to assist in the design of the earth station installation. This can be obtained from a manufacturer, with the obvious drawback that his advice may not be completely objective, or from an independent consultant. The customer must make this choice. In either case, it will be advantageous for him to understand the system design process sufficiently to carry on an effective dialogue with the consultant.

Equipment Suppliers

This section provides a brief sampling of some of the leading suppliers of earth station equipment. It is not a complete list, but it is a starting point for the development of a bidders' list that would be appropriate for a particular requirement.

Antenna Suppliers

Andrew Corp.
10500 W. 153d St.
Orland Park, IL 60462

General Instrument Corporation
Jerrold Division
2200 Byberry Road
Hatboro, PA 19040

Scientific-Atlanta, Inc.
One Technology Parkway
Box 105600
Atlanta, GA 30348

LNAs, LNRs, and Receivers

Aventek, Inc.
481 Cottonwood Drive
Milpitas, CA 95035-7492

General Instrument Corporation
Jerrold Division
2200 Byberry Road
Hatboro, PA 19040

LNR Communications, Inc.
180 Marcus Blvd.
Hauppage, NY 11788

Scientific-Atlanta, Inc.
One Technology Parkway
Box 105600
Atlanta, GA 30348

Scramblers, Descramblers
General Instrument Corporation
Jerrold Division
2200 Byberry Road
Hatboro, PA 19040

Exciters, HPAs
Harris Corporation
Box 1700
Melbourne, FL 32901

Scientific-Atlanta, Inc.
One Technology Parkway
Box 105600
Atlanta, GA 30348

Mobile Earth Stations
Satellite Transmission Systems Inc.
(Subsidary of California Microwave Inc.)
125 Kennedy Drive
Hauppage, NY 11788

Appendix 1

Conversion of FCC Downlink Power Density Limits to EIRP

The conversion of the FCC downlink power density limits (power per meter^2 per 4 kHz) to EIRP requires an assumption as to the distribution of carrier and sideband energy. For this purpose, it is reasonable to assume that an energy dispersal waveform (see Chapter 4) superimposed on the video signal results in a uniform distribution of energy over a 2 MHz band at the center of the transponder pass band.

On the basis of this assumption, the maximum permissible C-band EIRP in terms of the FCC limit is the following:

$$\text{EIRP}_{max}\text{dBw} = [\text{FCC limit}]\text{dBw/m}^2/\text{4kHz} + 163 + 10\log 2 \times 10^6/4 \times 10^3$$

Appendix 2
Antenna Directivity

The equations for calculating the half-power beam width (HPBW) and null beam width (angular separation of the nulls on either side of the main lobe) are the following:

$$\text{HPBW} = C_{HP}(\lambda/D)$$

$$\text{Null beam width} = C_{null}(\lambda/D)$$

where:

λ is the wavelength.

D is the antenna diameter.

C_{HP} and C_{null} are constants which depend on the uniformity of reflector illumination. For uniform illumination, $C_{HP} = 50°$ and $C_{null} = 114°$. For highly nonuniform illumination, $C_{HP} = 95°$ and $C_{null} = 280°$. For the examples in Table 6 which are typical of practical designs, $C_{HP} = 70°$ and $C_{null} = 170°$.

Appendix 3
Antenna Gain

The gain of an antenna with respect to an isotropic radiator expressed in decibels (dBi) is as follows:

$$\text{Gain} = 10 \log([4.1 \times 10^4][E]/[HPBW_1][HPBW_2]) \text{ dBi}$$

where:

E is the antenna efficiency. Antenna efficiencies of range from 50 to 85 percent. For the examples in Table 6, which are typical of practical designs, an efficiency of 70 percent was assumed.

$HPBW_1$ and $HPBW_2$ are the half-power beam widths in orthogonal planes in degrees. For a circular antenna with a symmetrical pattern, this factor becomes $[HPBW]^2$.

Appendix 4

Noise Temperature

The equation relating thermal noise power to temperature is the following:

$$\text{Noise power} = kBT$$

where:

k is the Boltzman constant = 1.38×10^{-23} Joules/K°/Hz

B is the bandwidth in Hertz

T is the absolute temperature in degrees Kelvin

$$T_{Kelvin} = T_{Centigrade} + 273°.$$

Appendix 5
Antenna Elevation and Azimuth Angles

To determine the elevation and azimuth angles of the path to a geosynchronous satellite, first calculate the angle, β, and the path length, L:

$$\beta = \cos^{-1}(\cos[\Delta \text{ long}]\cos[\text{lat}])$$

where [Δ long] is the difference between the longitudes of the earth station and the satellite and [lat] is the earth station latitude.

$$L = (18.2 - 5.4\cos[\beta]) \times 10^4 \text{ km}.$$

(L varies from 35,800 km with the satellite directly overhead to 39,550 km at an elevation angle of 10°.)

The elevation angle is then given by the equation:

$$\text{Elevation angle} = \cos^{-1}(4.22 \times 10^4/L)(\sin[\beta])$$

and the azimuth angle is given by the equation:

$$\text{Azimuth} = (180° + \tan^{-1}(\tan[\Delta \text{ long}]/\sin[\text{lat}])$$

Glossary

Ad hoc network   A television or radio network that is formed for a limited period of time to cover a specific event, for example, a major sports contest.

ADDS   Audio digital distribution service. Also called DATS (digital audio transmission service). A system for the distribution of radio programs by satellite, using a digital signal.

AKM   Apogee kick motor. An explosive charge in the satellite that provides the energy for the final stage of the launch and places the satellite in the geosynchronous orbit.

B-MAC   Multiple analog components, type B. A format for the transmission of color television programs with three separate signals. This is in contrast to the multiplexed NTSC and PAL systems that transmit the color information with a high-frequency subcarrier.

Baseband   The voice, video, audio, or data signal transmitted by a communication system.

Beacon   A small transmitter installed on a satellite that transmits continuously. It is used for aiming earth station antennas and automatic power level control for the Ku-band uplinks.

Bus   See Space vehicle.

C band   A portion of the radio frequency spectrum. Communication satellites are allocated C-band frequencies from 5.925 GHz to 6.425 GHz for uplinks and 3.700 GHz to 4.200 GHz for downlinks.

Cassegrain antenna   A dual reflector antenna in which the subreflector is convex toward the main reflector.

CCIR   Comite Consultatif de Radio Communication. An international standards body for all forms of radio communication.

Cleared site   An earth station site that is free from interference from microwave systems and other earth stations.

Communications Satellite Act of 1962   An Act of Congress that delegated authority for the regulation of communications satellites to the Federal Communications Commission (FCC).

Comsat   Communications Satellite Corporation. A private company formed in 1962 pursuant to the Communications Satellite Act. It is the chosen instrument of the United States to hold this country's interest in Intelsat, q.v., and to operate the United States ground segment of the Intelsat system.

Coordination   A term applied to the site selection process required by the FCC to make sure that an earth station on the site will neither cause nor receive excessive interference from existing stations.

Cross-Strapping   Interconnecting C-band and Ku-band transponders on a hybrid satellite so that the uplink is in one band and the downlink in the other.

DATS   Digital audio transmission service. Also called ADDS (digital audio transmission service). A system for the distribution of radio programs by satellite, using digital signals.

DBS   Direct broadcast by satellite. A system of television program distribution in which programs are transmitted direct from a satellite to homes. In its specialized usage, the term applies to very high-power satellites that permit reception with very small antennas.

DBS Authorization Center   A facility established by the General Instrument Corporation in La Jolla, California, to provide homesats with authorization codes to descramble programs.

Deemphasis   See Preemphasis.

Downconverter   A component of a satellite receiver that translates the high-frequency carriers of a group of channels (or of a single channel) downward to an intermediate frequency for amplification.

Downlink   A radio circuit from satellite to earth station.

Dual-reflector antenna   An antenna consisting of a parabolic reflector that is illuminated by reflected energy from a subreflector.

Earth station   An installation for the transmission and/or reception of radio signals to and from a satellite.

EIA   Electronic Industries Association. The major trade association of electronic manufacturers in the United States.

EIRP   Effective isotropic radiated power. The ratio, expressed in decibels, between the radiated power density and the power density radiated by a 1-watt transmitter from an isoptropic radiator.

Elevation-over-azimuth mount   An antenna mounting system in which the elevation and azimuth angles are adjusted independently.

ENG   Electronic news gathering. A system of television journalism in which live pictures of news events are recorded on tape or transmitted from the scene to the station by electronic means, usually microwave or satellite. See SNG.

Fade margin   The ratio of the carrier-to-noise level in a receiver under normal operating conditions to the receiver threshold.

Faraday effect   The rotation of the plane of polarization of a radio wave when subjected to a strong magnetic field.

Ferrite polarizer   A device for rotating the plane of polarization of a radio wave by the use of the Faraday effect.

Field effect transistor   A type of transistor with a very low noise temperature, frequently used in the input stage of satellite receivers.

FM improvement factor The ratio, expressed in decibels, between the signal-to-noise ratio in the demodulated output of a frequency modulated carrier and the carrier-to-noise-ratio.

Footprint The area on the surface of the earth that receives a signal of useful strength from a satellite.

Frequency Reuse Doubling the capacity of a satellite by the use of cross polarization.

FSS Fixed service satellites. Satellites used for communication services that are neither broadcast nor mobile.

G/T In an earth station receiving system, the ratio of the antenna gain to the receiver noise temperature. It is a fundamental figure of merit for a receiving system.

Geosynchronous orbit A circle 22,300 miles above the earth in the equatorial plane. The centrifugal force and the force of gravity on an object in this orbit are exactly balanced when it rotates in synchronism with the earth.

Geosynchronous satellites Satellites located in the geosynchronous orbit.

Gregorian antenna A dual-reflector antenna in which the subreflector is concave toward the main reflector.

Gregux antenna A dual-reflector antenna in which the subreflector is a concave ring.

Ground segment The terrestrial portion of a satellite communication system, that is, its earth stations.

Half-power beam width (HPBW) The angle between the points on the main lobe of an antenna beam where the radiated power density is one-half that at the peak of the lobe.

Homesat A low-cost TVRO, q.v., designed for home use.

HPA High-powered amplifier. The final amplifier in an uplink transmitter.

Hybrid satellites Satellites having both K-band and C-band transponders.

Intelsat International Telecommunications Satellite Consortium. An international organization that owns and operates the satellites used for international satellite communication services.

ISDN Integrated service digital network. A communication system for the transmission of voice, audio, video, and data traffic on a high-speed digital stream.

Isotropic radiator A hypothetical antenna that radiates equally in all directions.

ITU International Telecommunication Union. An international regulatory body that has been delegated the authority to establish policies in international communication matters. This includes the allocation of orbital slots to the countries of the world.

Ku band A portion of the radio frequency spectrum. Communication satellites are allocated Ku-band frequencies from 14.0 MHz to 14.5 MHz for uplinks and

12.2 to 12.7 GHZ for downlinks. DBS satellites are allocated Ku-band frequencies from 17.3 to 17.8 GHz for uplinks and 12.2 to 12.7 GHz for downlinks.

LNA   Low-noise amplifier. The initial stage in an earth station receiver designed to generate very little noise.

LNB   Low-noise block converter. A component that combines the functions of a low-noise amplifier and a downconverter.

Look angle   The elevation angle, that is, the angle above the horizontal, of a satellite as seen from a point on the earth.

Loop test translator   A unit of test equipment that simulates the uplink/downlink frequency conversion in a satellite so that a complete earth station can be tested without using a satellite.

Magnetic variation   The angle at a point on the earth's surface between magnetic north and true north.

Modulation index   In an FM system, the ratio of the frequency deviation to the highest modulating frequency.

Noise   Broadly, a term used to describe any unwanted electrical disturbance in a communication signal. Specifically, a term used to describe a random disturbance caused by thermal effects (thermal noise) or similar in nature to thermal noise but not of thermal origin. The visual effect of thermal noise in a video signal is a *snowy* picture; in a radio signal it produces *hiss*.

Noise temperature   A criterion for specifying the electrical noise power generated by a system component or a complete system. It is defined as the temperature at which a hot object would generate or emit the same amount of thermal noise power.

Nonionizing radiation   Radiation, for example, satellite transmissions, that does not ionize molecules that are exposed to it. This distinguishes it from ionizing radiation such as X-rays.

NRTC   The National Rural Telecommunication Co-op. A third party (controlled neither by cable TV operators nor program suppliers) organization that was formed to receive orders from the public for scrambled pay-TV services and to inform the appropriate authorization center.

NTSC   National Television Systems Committee. An *ad hoc* committee convened from time to time by the National Association of Broadcasters and the Electronic Industries Association to study and recommend television standards and specifications.

NTSC color system   The system developed by the NTSC in 1952 and later approved by the FCC for color broadcasting in the United States.

Open skies policy   A policy announced in an Order issued by the FCC in 1972 stating that any legally and financially qualified U.S. citizen or corporation could apply for and be granted an orbital slot.

Orbital slot   A fixed location in the geosynchronous orbit that is assigned by regulatory authorities to a specific satellite.

PAL color system   Phase alternating lines. A color system, which is a variant of the NTSC system, that is used widely in Europe.

Parametric amplifier   A type of amplifier with a very low noise temperature, frequently used in the input stage of satellite receivers.

Payload   The communication system components in a satellite.

Pointing accuracy   The rms angular deflection of an antenna, measured over a period of time, at stated levels of wind velocity.

Polar mount   An antenna mounting system in which the antenna is rotated around an axis that is parallel to the earth's axis.

Preemphasis   Enhanced amplification of the higher frequency components of video and audio signals before transmission. The complimentary deemphasis at the receiver improves the signal-to-noise ratio.

Preemptible service   Unprotected satellite service that can be preempted by services of higher grade.

Prime orbital arcs   The portion of the geosynchronous orbit in which the look angle, q.v., exceeds 5° for C band or 10° for Ku band in the desired service area.

Prime-focus-feed antenna   An antenna consisting of a parabolic reflector that is illuminated directly with a feed horn located at its focus.

Program Syndicators   Independent entrepreneurs who obtain the rights for television programs for resale to broadcasters.

Protected service   Satellite service provided with back-up facilities that can be used to continue service in the event of a satellite or transponder failure.

PSK   Phase shift keying. A modulation mode used for the transmission of digital signals.

RARC   Regional Administrative Radio Conference. A conference at which the policy making functions of the ITU for matters of interest to a single region are exercised.

Receiver threshold   The carrier-to-noise ratio below which the signal-to-noise ratio of the receiver output declines more rapidly than the carrier-to-noise ratio.

Resale carrier   An organization that leases communication facilities on a long-term wholesale basis and subleases them on a short-term retail basis.

Scrambling   The deliberate distortion of video and/or audio signals to prevent their reception by unauthorized viewers.

Sideband energy dispersal   The spreading of the sideband energy from a frequency modulated carrier over an area of the spectrum to reduce the energy density expressed in watts per kHz of bandwidth.

Site coordinator   A consultant who performs the earth station site coordination functions required by the FCC for new sites.

SNG   Satellite news gathering. A form of ENG (q.v.) in which satellite is the communication medium.

Solar eclipse   The time interval when the earth comes between the sun and the satellite, thus making solar energy unavailable.

Space segment The portion of a satellite communication system located in space, that is, the satellite.

Space shuttle A reusable launch vehicle that elevates the satellite to an elevation of about 200 miles and then returns to earth.

Space vehicle The housing and electrical and mechanical systems of the satellite that provide a working environment for the payload while maintaining it in its orbital slot, sometimes called the bus.

Specialized network A network with programming that is regional or that has specialized content.

Spectrum analyzer A type of test equipment that displays the amplitude of signals in an entire region of the spectrum.

Spin-stabilized satellites Satellites whose orientation is controlled by spinning the entire housing.

SSPA Sold state power amplifier. A high-power amplifier employing solid state components.

Station keeping The maintenance of a satellite in its assigned orbital slot and in the proper orientation.

Sun outage The blocking of a downlink signal as the result of excessive solar noise in the receiver when the satellite and sun are in line as seen from the earth station.

Teleport A major earth station installation with extensive facilities for accessing a number of satellites and with terrestrial interconnections.

Three-axis-stabilized satellites Satellites whose attitude is controlled by internal gyroscopes. See spin-stabilized satellites.

Torus antenna A multiple-beam antenna in which the reflecting surface is circular in the plane of the feed horns and parabolic in the plane orthogonal to the feed horns.

Transponder An electronic component of a satellite that shifts the frequency of an uplink signal and amplifies it for retransmission to the earth in a downlink.

TT&C An earth station used for tracking, telemetry, and control of satellites in orbit.

TVRO Television, receive only. An earth station designed for downlink service only.

TWTA Travelling wave tube amplifier. A device for generating comparatively large amounts of power in the microwave regions of the spectrum.

Unprotected service Satellite service with no backup facilities. Unprotected service, however, cannot be preempted.

Upconverter A component of an uplink earth station that translates the intermediate-frequency carriers of a group of channels (or of a single channel) upward to the final carrier frequency.

Uplink A radio circuit from earth station to satellite.

Videocipher[R] The trade name for a scrambling system manufactured by the General Instrument Corporation. It has become the de facto scrambling standard for cable TV programs.

VSAT   Very small aperture terminals. Small earth stations, antenna diameter approximately 1.8 meters, used for transmission of digital signals at a 56 or 64 kbs rate. Can be used for teleconferencing.

WARC   World Administrative Radio Conference. A conference at which the policy making functions of the ITU for matters of worldwide interest are exercised.

Weighted noise   The noise power in a circuit or system calculated by weighting its frequency components in proportion to their visibility for video signals and audibility for audio signals.

Notes

[R] Videocipher is a registered trademark of the General Instrument Corporation.

Bibliography

Reference Books

Benson, K. Blair, *Television Engineering Handbook*. New York: McGraw-Hill, New York, 1986.

British Broadcasting Corporation. *Handbook*. London: BBC, annual.

Broadcasting/Cable Yearbook. Washington, D.C.: Broadcast Publications, Inc., annual.

Fink, Donald G., and Donald Christiansen. *Electronic Engineers Handbook*. 3d ed. New York: McGraw-Hill, 1989.

Inglis, Andrew F. *Electronic Communications Handbook*. New York: McGraw-Hill, 1988.

Television & Cable Factbook: Cable & Services Volume. Washington, D.C.: Warren Publishing Inc., annual.

Technical Books

Barlett, Eugene R. *Cable Television Technology and Operations*. New York: McGraw-Hill, 1990.

Ha, Tri T. *Digital Satellite Communications,* 2nd ed. New York: McGraw-Hill, 1990.

Magnant, Robert S. *Domestic Satellites: An FCC Giant Step Toward Competitive Communications*. Boulder, CO:Westview Press, 1977.

Miya, K. *Satellite Communications Technology*. Tokyo: KDD Engineering and Consulting, Inc., 1981.

Watkinson, John, *The Art of Digital Video*. London, Boston: Focal Press, 1990.

Standards and Recommendations

CCIR Recommendations and Reports, CCIR, Plen. Assy., Geneva, 1982, Vol. XI, Broadcasting Service (Television), ITU, Geneva, 1982; especially:

CCIR Recommendation 353, Analog Systems

CCIR Recommendation 567, Analog television

CCIR Recommendation 579, Availability

CCITT Recommendation G222, Analog systems

CCIR Recommendation 355, Genral Sharing [spectrum]

CCIR Recommendation 356, Interference from FS
   [fixed service, i.e., microwave] into FSS FM systems

CCIR Recommendation 357, Interference from FSS into FS analog angle-modulated systems

EIA Standards, EIA Engineering Department, Washington, D.C.; especially:

EIA/TIA-250-C Video performance standards

EIA/TIA-250-C Audio performance standards

Regulatory

FCC Rules and Regulations, Satellite Communications, Title 47, Part 25 of the Code of Federal Regulations. Government Printing Office, Washington, D.C.

Index

Uplink frequencies, 28-29
Usage classifications, 28

Video Cipher II and II Plus scramblers, 23,
    45-46
VideoPal, 46
Video performance standards, 56-58
Video preemphasis, 13, 47
VLSI (very-large-scale-integrated) circuit, 46
Voice communications, 29
VSAT (very small aperture terminal) net-
    works, 25

Waveform spreading, 53
Westar satellites, 18
Western Union, 16, 18
Wind loading, 43
World Administrative Radio Conference
    (WARC), 36, 72

Zoning ordinances, 59